SpringerBriefs in Applied Sciences and Technology

PoliMI SpringerBriefs

Series Editors

Barbara Pernici, DEIB, Politecnico di Milano, Milano, Italy

Stefano Della Torre, DABC, Politecnico di Milano, Milano, Italy

Bianca M. Colosimo, DMEC, Politecnico di Milano, Milano, Italy

Tiziano Faravelli, DCHEM, Politecnico di Milano, Milano, Italy

Roberto Paolucci, DICA, Politecnico di Milano, Milano, Italy

Silvia Piardi, Design, Politecnico di Milano, Milano, Italy

Gabriele Pasqui, DASTU, Politecnico di Milano, Milano, Italy

Springer, in cooperation with Politecnico di Milano, publishes the PoliMI Springer-Briefs, concise summaries of cutting-edge research and practical applications across a wide spectrum of fields. Featuring compact volumes of 50 to 125 (150 as a maximum) pages, the series covers a range of contents from professional to academic in the following research areas carried out at Politecnico:

- Aerospace Engineering
- Bioengineering
- Electrical Engineering
- Energy and Nuclear Science and Technology
- Environmental and Infrastructure Engineering
- Industrial Chemistry and Chemical Engineering
- Information Technology
- Management, Economics and Industrial Engineering
- Materials Engineering
- Mathematical Models and Methods in Engineering
- Mechanical Engineering
- Structural Seismic and Geotechnical Engineering
- Built Environment and Construction Engineering
- Physics
- Design and Technologies
- Urban Planning, Design, and Policy

Ingrid Paoletti · Massimiliano Nastri

Technology of the Multi-Layer Façade Systems

Executive Design, Construction Methodologies and Experimental Applications

Ingrid Paoletti [iD]
Department of Architecture, Built
Environment and Construction Engineering
(ABC)
Politecnico di Milano
Milan, Italy

Massimiliano Nastri
Department of Architecture, Built
Environment and Construction Engineering
(ABC)
Politecnico di Milano
Milan, Italy

ISSN 2191-530X	ISSN 2191-5318 (electronic)
SpringerBriefs in Applied Sciences and Technology
ISSN 2282-2577	ISSN 2282-2585 (electronic)
PoliMI SpringerBriefs
ISBN 978-3-032-04766-3	ISBN 978-3-032-04767-0 (eBook)
https://doi.org/10.1007/978-3-032-04767-0

© The Editor(s) (if applicable) and The Author(s), under exclusive license to Springer Nature Switzerland AG 2025

This work is subject to copyright. All rights are solely and exclusively licensed by the Publisher, whether the whole or part of the material is concerned, specifically the rights of translation, reprinting, reuse of illustrations, recitation, broadcasting, reproduction on microfilms or in any other physical way, and transmission or information storage and retrieval, electronic adaptation, computer software, or by similar or dissimilar methodology now known or hereafter developed.
The use of general descriptive names, registered names, trademarks, service marks, etc. in this publication does not imply, even in the absence of a specific statement, that such names are exempt from the relevant protective laws and regulations and therefore free for general use.
The publisher, the authors and the editors are safe to assume that the advice and information in this book are believed to be true and accurate at the date of publication. Neither the publisher nor the authors or the editors give a warranty, expressed or implied, with respect to the material contained herein or for any errors or omissions that may have been made. The publisher remains neutral with regard to jurisdictional claims in published maps and institutional affiliations.

This Springer imprint is published by the registered company Springer Nature Switzerland AG
The registered company address is: Gewerbestrasse 11, 6330 Cham, Switzerland

If disposing of this product, please recycle the paper.

Preface

The study, developed inside the Poli-Façades sector of the experimental laboratory *Material Balance Research* within the Architecture, Built Environment and Construction Engineering—ABC Department at the Politecnico di Milano, examines the potentialities of the advanced envelope systems with respect to the principles of sustainability aimed at environmental interaction procedures, directed at the calibration of the climatic stresses, the containment of dissipation and the aid of renewable energy sources. The advanced envelope systems under investigation, expressed as an evolution of the façade systems (as an evolved configuration of the curtain wall, within the geometric and material continuity of the light façades), are carried out with respect to the contemporary buildings of an experimental nature (with especially tertiary and commercial uses), in customized forms and in analogy with the notion of the machine envelope. The analysis considers the advanced envelope systems as mechanical bodies, active diaphragms and membranes that promote or hinder exchanges (luminous, thermal, aeriform and acoustic) with the external environment, playing a role in energy regulation. Furthermore, the analysis observes the construction and use of techniques, components and functional devices aimed at both the reduction of losses or the accumulation of heat (resulting from radiation), and the dynamic selection of sunlight and the calibration of natural light.

Therefore, the study of the envelope makes explicit the physical, material and performance contents according to the criteria of action towards both energy and environmental conditions and ergonomic conditions through the procedures of reflection, capture and diffusion of the stresses external or internal to the built spaces. The research focuses on the passive type of transformation, aimed at accumulating and distributing the energy produced by solar radiation without the use of plant equipment and without the contribution of devices (in the form of collectors) aimed at integrating and conveying heat, natural light or convection phenomena related to the airflows. The applied technologies and, in particular, the components and devices of the envelope are assumed with respect to the processes of eco-efficient interaction and permeability towards thermo-hygrometric, light and air stresses (determining the criteria of energy and environmental control of a selective and dynamic type), with the possibility of regulating the flows and conveying them in the overall functioning of the construction. In this experimental scenario, the advanced envelope systems are studied as mediating and reacting apparatuses towards external loads, in accordance with settlement needs and requirements (as an environmentally conscious design activity). The methodological structure of the study intends to examine the advanced envelope systems towards the principles of the environmental and energy sustainability through:

- the drafting of an explanatory and operational reference contribution, in the absence of cognitive and instrumental supports (within the international scientific and academic literature) capable of exposing the typological, functional and constructive configuration of the systems under examination on the basis of the investigation around the experimental applications towards new buildings;
- the use of the executive re-elaboration of the systems as a technical tool for understanding and knowledge of the functional modes and sustainable procedures;
- the drafting of a technical guide to understanding the construction and functional modes, focusing the analysis on the design and construction aspects;
- the definition of the main functional typologies, established on the basis of the analysis conducted on the experimental case studies, with the aim of achieving at a categorization of the performance and executive potentialities;
- the possibilities of technology transfer towards common construction, with the aim of increasing environmental and energy sustainability also through applications on the external perimeter enclosures of the existing buildings;
- the possibilities of technology transfer through the use of standard and non-customized products, so as to increase the diffusion also in the production and construction of mass-produced systems and components.

Ingrid Paoletti
Scientific Director of the Material
Balance Research Group, Department
of Architecture, Built Environment
and Construction Engineering (ABC)
Politecnico di Milano
Milan, Italy
ingrid.paoletti@polimi.it

Massimiliano Nastri
Material Balance Research Group,
Scientific Coordinator of Poli-Façades,
Department of Architecture, Built
Environment and Construction
Engineering (ABC)
Politecnico di Milano
Milan, Italy
massimiliano.nastri@polimi.it
https://www.materialbalance.polimi.it

Competing Interests The authors have no competing interests to declare that are relevant to the content of this manuscript.

Contents

1 The Environmental Constitution of the Multi-layer Façade Systems 1
    1.1 The Constructive, Environmental and Interactive Composition of the Multi-functional Layered Envelopes 2
    1.2 The Functional Development of the Multi-layer Façade Systems: A Scientific Analysis 6
    References 15

2 The Functional and Executive Typologies of the Multi-layer Façade Systems 17
    2.1 The Methodology of Identification and Analytical Classification of the Multi-layer Façade Systems 18
    2.2 The Multi-layer Functional and Executive Typology of the Box-Windows Façade Systems 19
    2.3 The Multi-layer Functional and Executive Typology of the Multistorey Façade Systems 27
    2.4 The Multi-layer Functional and Executive Typology of the Shaft-Box Façade Systems 35
    2.5 The Multi-layer Functional and Executive Typology of the Corridor Façade Systems 39
    2.6 The Scientific and Executive Perspectives: Potential and Criticality of Multi-layer Façade Systems Development and Application 53
    References 57

3 The Technical Hybridization of the Multi-layer Façade Systems 59
    3.1 The Executive and Functional Constitution of the Integrated Façade Systems 60
    3.2 The Executive and Functional Constitution of the Combined Performance Façade Systems 66

**4 The Technological Requalification by the Multi-layer Façade
Systems** 81
   4.1 The Executive and Functional Constitution
of the Environmental and Interactive Façade Systems 82
   4.2 The Architectural and Performance Reconfiguration
by Technological Substitution of Envelope Systems 94

**5 The Technology Transfer of the Multi-layer Façade Systems
by Mass Production Components** 113
   5.1 The Technology Transfer Methods by the Transformation
of Experimental and Custom Processes 114
   5.2 The Composition of Functional and Executive Principles
by the Use of Technical and Mass Production Elements 121

Chapter 1
The Environmental Constitution of the Multi-layer Façade Systems

Abstract The study focuses on the multi-layer envelope systems (understood as multiple-skin façades), whereby the combination of the planar surfaces generates equipment with greenhouse effect, chimney effect and natural ventilation (in the form of double-skin façades): then, performance is concretized on the basis of treatment practices (thermal, chemical and surface), layering and coating (acting on the transmission of visible, solar and thermal radiation, especially with the reference to the infrared spectral field) and deposition on the perimeter enclosures. With respect to the enclosure elements, the application focuses on the physic of the glazed, combined and multi-layered surfaces, which experimental research tends to transform into dense, intelligent systems interfaces: the main materials constituting the external surfaces are composed according to their processes of change from stable entities to designable entities according to a particular performance program. In this case, the application of the materials of the envelope, in the form of projectable entities, is structured with respect to the outcomes of the solutions in which functions tend to become complex (in a controlled and managed manner) and articulated with each other, realizing multiple performances through the correlation of different agents and layers. The advanced envelope systems are specified in the constitution of integrated functional components with the objective of receiving, guiding and selecting the environmental loads to realize calibrated ergonomic conditions in the built spaces. For this, the systems are endowed with engineering performances (such as multiple environmental performances), are articulated in the form of environmentally responsive walls (capable of actively responding to environmental stresses through perceptual and organic contact with the climatic conditions) and as engineered walls (such as equipment that can be operated by mechanical devices), aimed at regulating the transmission of heat, light and natural ventilation, along with the attenuation of wind and acoustic loads.

Keywords Multiple-skin façades · Integrated façades · Envelope environmental functioning · Envelope dynamic interaction · Passive solar systems · Parietodynamic transfer processes · Climate interaction systems

1.1 The Constructive, Environmental and Interactive Composition of the Multi-functional Layered Envelopes

The multi-layer façade systems (such as multiple-skin façades), as in the form of double-skin façades, are determined by the cavity between the inner curtain and the outer screen, for thermal and acoustic insulation, for ventilation and for the insertion of functional devices (such as sunscreens) and, also, plant ducts. The cavity between the two enclosures constitutes a ventilated cavity that can be used according to certain modes of functioning aimed at controlling external climatic and environmental stresses in order to regulate the conditions of the interior spaces (Schittich 2001; Eren and Erturan 2013). In this regard, the environmental and adaptive strategy processes the envelope systems according to their metabolic efficacy and instinctual reactive capacity, configuring them as *intelligent skins* endowed with automatic performance (by means of functional criteria of autonomous regulation) and membranes defined as *biological skins* (active against external agents by means of sensors and protective devices). In particular, the use of elements with dynamic and reactive behaviour assumes the use of the surfaces for the control of the solar radiation, consisting of filtering or screening sections capable of modulating their transparency according to the level and distribution of the natural luminosity in the interior spaces (Lovell 2010). The development of the systems thus exposes the *techno-organic* qualities directed at the functioning of the vertically developed buildings, through the interpretation and assimilation of the environmental conditions in a manner combined with the use of the evolved techniques (in *organi-tech* form): the study includes the experimentation around the artificial systems integrated with the natural systems, as tools for the accumulation, conveyance, protection and calibration of the passive energies that can provide buildings with forms of heating, air conditioning and ventilation (Abrantes et al. 2017).

The multiple-skin façades are realized as an apparatus of mediation and reaction towards the environmental loads, according to the needs of well-being and reduction of the energy consumption. In general, the systems offer the functioning in the form of a passive solar system, assuming the use and accumulation of solar radiation for the regulation of the indoor thermal comfort conditions, and the functioning for the capture and input of the airflows. The calibration of the solar radiation is integrated with the use of the shading devices in order to achieve the diffuse lighting conditions in interior spaces. In addition, the application of the double-glazing surface reduces the thermal losses from the interior spaces by decreasing the speed of the airflow in contact with the inner curtain and increasing thermal insulation (Syed 2012; Schmid et al. 2018; Paoletti and Nastri 2023).

The system foresees that the ventilated cavity performs various integrated functions (for the definition of complex mechanisms of dynamic interaction with the external climatic conditions), both permanent (e.g. for the increase of the thermal inertia and the acoustic insulation related to the internal curtain) and temporary (e.g. for the cooling of the same spaces during the periods of high temperature). The multiple-skin façades constitution provides the use of a screen (or second skin, in general, made of glass) outside the vertical enclosure, with the aim of optimizing the functions allowed in the cavity: this is through the additional application of the glass enclosure in front of the curtain wall or the external building curtain (in general, equipped with openable frames), in the form of a ventilated cavity that can be used according to certain modes of operation aimed at controlling the external climatic and environmental loads to regulate the conditions of the internal spaces (Hausladen et al. 2008; Hamid 2012; Stazi 2019) (Fig. 1.1).

The study detects the prospects of the dynamic interaction between the multiple-skin façades and the external environment, observing:

- the criteria aimed at realizing built spaces in a stable and balanced manner, with the possibility of transmitting, modifying or rejecting the climatic stresses;
- the development of an interchange tool for the ability to respond to the external loads through the different functional levels and the use of means of regulation to manipulate the interactions with the environment (Figs. 1.2, 1.3 and 1.4).

The paradigms of sustainability, the propensity (ethical, operational and legislative) to contain energy consumption and reduce the use of non-renewable sources

Fig. 1.1 Renzo Piano Building Workshop, *Float Building*, Düsseldorf. Construction of the main perimeter curtain wall, the horizontal overhanging section (by means of the structural brackets connected to the framing) and the external glazed screen. © Courtesy of RPBW, Carla Baumann

find in contemporary experimentation around the envelope (and its plant and environmental integrations) the potentialities and possible performance indications to be examined and transferred also within the widespread production. The mechanical conception of the envelope is combined with a direct approach to the ecological and energy issues, non-invasive insertion into the site, use of alternative sources and passive indoor environmental climate control strategies. According to this orientation, the physical configuration is integrated with the organization of the spaces, services and management devices, considering the development of the envelope as an optimized system in order to manipulate the control modes for the direct energy gains, thermal protection and natural ventilation (Daniels 1994; Slessor 1997; Knaack et al. 2007).

The façade system, as an apparatus for mediating and reacting to the external inputs, focuses, on the one hand, on the adaptation and control of the thermal and energy exchanges between the internal micro-environment and the external macro-environment, recognizing this area as a catalyzing factor in the type-technological evolution of the architecture; on the other hand, on the appropriate solutions to optimize the same exchanges, as well as on the expressive content (Altomonte 2004; Herzog et al. 2004; Kim et al. 2012).

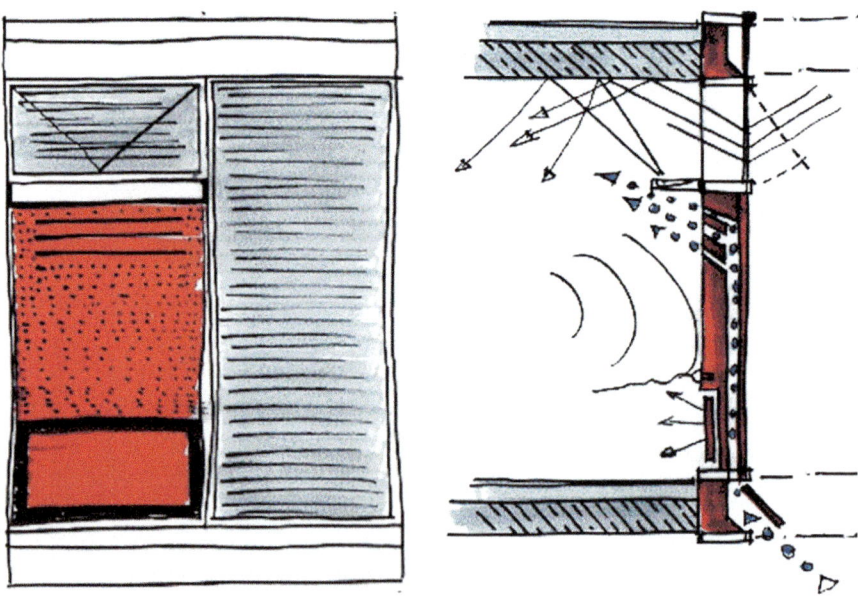

Fig. 1.2 Gatermann + Schossig, *Capricorn Haus*, Düsseldorf. Climatization takes place through the activation of the inertial mass of the horizontal structures, while the peripheral ventilation and air cooling are included in the façade system; in this way, the hot or cold air is introduced into the interior spaces through an air-conditioning system equipped with a concrete core (positioned on the ceiling), which generates a high mass of thermal inertia. © Courtesy of Schüco

Fig. 1.3 Gatermann + Schossig, *Capricorn Haus*, Düsseldorf. Division of the horizontal and vertical type that produces the double-skin façade type consisting of units that are independent of each other, in which the cavity is divided by separating elements, capable of generating a continuous ventilated chamber. © Courtesy of Schüco

Fig. 1.4 Gatermann + Schossig, *Capricorn Haus*, Düsseldorf. Envelope modulation by the use of the unit façade system considering the air exchange due to the thermal compensation effect; the ventilation slots are located at the base and the coated end. © Courtesy of Schüco

1.2 The Functional Development of the Multi-layer Façade Systems: A Scientific Analysis

The sustainable methods of the multiple-skin façade systems are analyzed as a mediating and reacting apparatus, specifying the sensitive type of functioning that acts with the adaptive and control capacities, according to the needs of well-being and reduction of the energy consumption. These systems are intended as programmable surfaces, capable of interpreting the functions and needs of the users in an *eco-efficient*, selective and multi-purpose form, with respect to the control of the temperature, humidity and ventilation levels, perception towards the outside and lighting levels. The technical and typological definition of these systems can be calibrated with respect to the activities inside the built spaces, referring to the main external environmental and micro-climatic parameters (i.e., the intensity of the solar radiation and its distribution) and the internal ones (such as the temperature of the air and the perimeter vertical curtains, relative humidity, air speed and its quality) (Aksamija 2013; Mohsen 2020; Gunawardena and Mendis 2022) (Figs. 1.5 and 1.6).

The composition of the multiple-skin façades considers the functioning in the form of a passive solar system (in general, through the capture of the radiant energy, the reduction of heat dispersion, the possibility of heat accumulation in the form of a thermo-insulating air chamber and, therefore, of heating the air section by the "greenhouse effect", the increase in lighting performance), based above all on the thermo-building principles related to the control of the airflows in the cavity (by means of the parietodynamic transfer processes). In this regard, the systems assume the use and accumulation of the solar radiation for the regulation of the indoor thermal comfort conditions, together with the possibility of a reduction in the use of the heating systems. The functioning in the form of a system for capturing and injecting the airflows (whereby the amount of air exchanged between the external environment and the cavity depends on the temperature gradient, wind pressure and the size of the ventilation slots) considers, in general, the passive type. This typology captures the convective flows close to the façade plane and introduces them into the cavity between the internal and external enclosures, conducting the convective flows by the "chimney effect" in an upward direction until they reach the internal spaces (considering the possibility of reducing the use of the air-conditioning systems for cooling). Moreover, the systems of the passive type are characterized by the absence of electro-commanded equipment to generate the aeration of the convective flows in the cavity between the internal and external enclosures (such as, for example, the fans for the conduction of airflows) (Nastri 2021). The calibration of the solar radiation, in an integrated manner with the use of the shading or diffusing devices, is directed to obtain diffused lighting conditions in the interior spaces (capable of limiting the use of the artificial light). These devices, placed in the cavity (protected from the atmospheric pollution and bad weather), keep the heat absorbed by the solar radiation outside the built spaces and determine the accumulation of the heat aimed at increasing the temperature of the air in the cavity, the flow of which is directed upwards, until it is expelled (through the ventilation devices) (Yonghuort

1.2 The Functional Development of the Multi-layer Façade Systems … 7

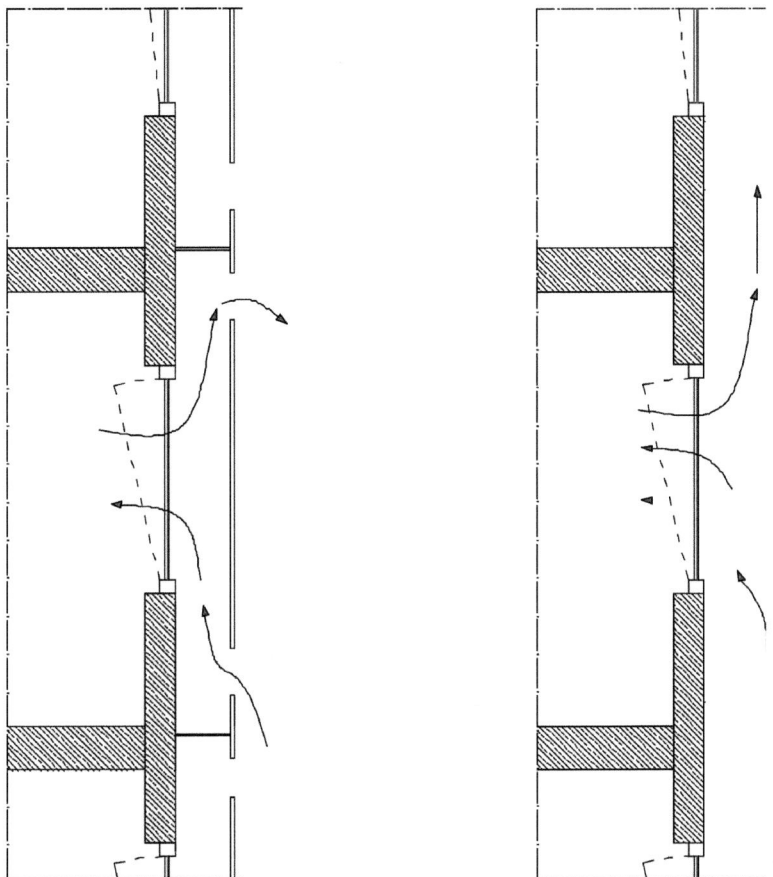

Fig. 1.5 Activation of the airflows in the cavity determined by the external screen, detecting the calibration of the air changes compared to the traditional surface with window opening

and Mohd 2022). This application allows the accumulation of the heat outside the built space, reducing the air conditioning loads and the need for cooling. In addition, the calibration of the solar radiation can be specified by the use of the devices capable of transmitting, reflecting and diffusing the natural lighting (Fig. 1.7).

The application of the double-glazed surface makes it possible to reduce the thermal losses from the internal spaces, by reducing the speed of the airflow in contact with the internal curtain, increasing the thermal insulation: therefore, the reduction in thermal transmission makes it possible to maintain the glass surfaces at a temperature close to the values of the average internal temperature, so as to make the adjoining spaces more comfortable. The change of the air inside the cavity increases in direct proportion to the solar radiation, since the airflows that lap the façade are heated by the elements that make it up (the glazed panels and metal profiles): therefore, the convective circulation and the amount of the evacuated heat

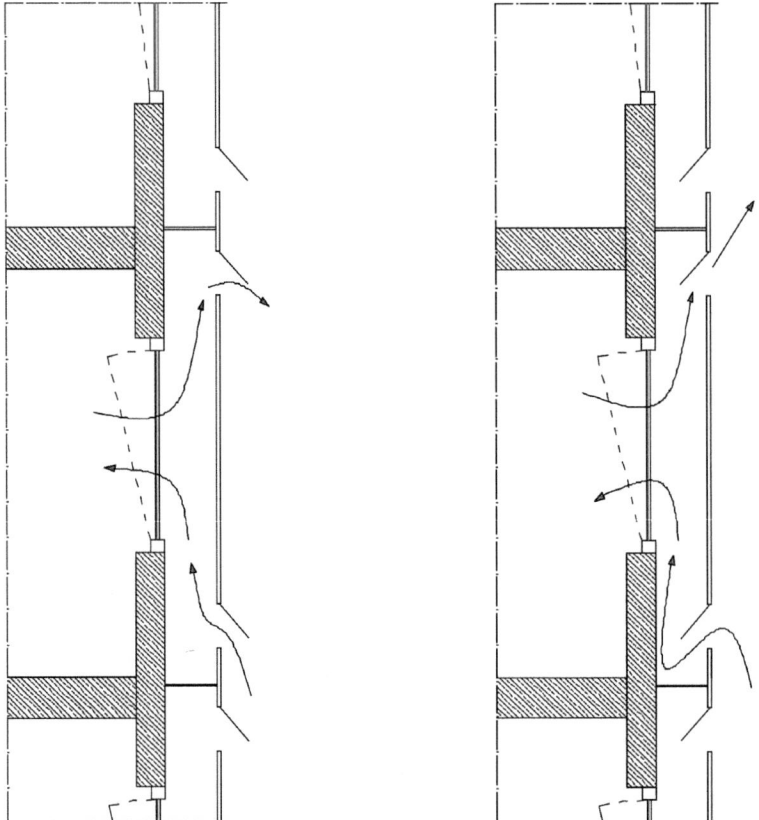

Fig. 1.6 Activation of the airflows in the cavity determined by the external screen, detecting the contribution of the adjustable openings in the cavity to convey incoming and outgoing airflows

increase according to the intensity of the solar radiation, so that the interior spaces are ventilated even in difficult climatic conditions (Daniels 2003; Boswell 2013). During the winter season, the cavity inside the multiple-skin façades systems acts as a passive heating device through the accumulation of the heat (due to the solar radiation), sheltering the internal surface from the effects of the low temperatures and improving the thermal insulation of the curtain wall (Fig. 1.8).

The solar shading devices (in an adjustable form) are placed inside the cavity and, therefore, protected from the atmospheric agents and the external pollutants: these devices reduce the heat input according to the external temperature and the solar radiation conditions, being particularly effective when the external temperature is lower than the temperature of the internal spaces and for the low values of total radiation. The external screen, in the form of a totally or partially transparent façade, is made up of monolithic tempered or double-glazed glass panels: this screen, in addition to

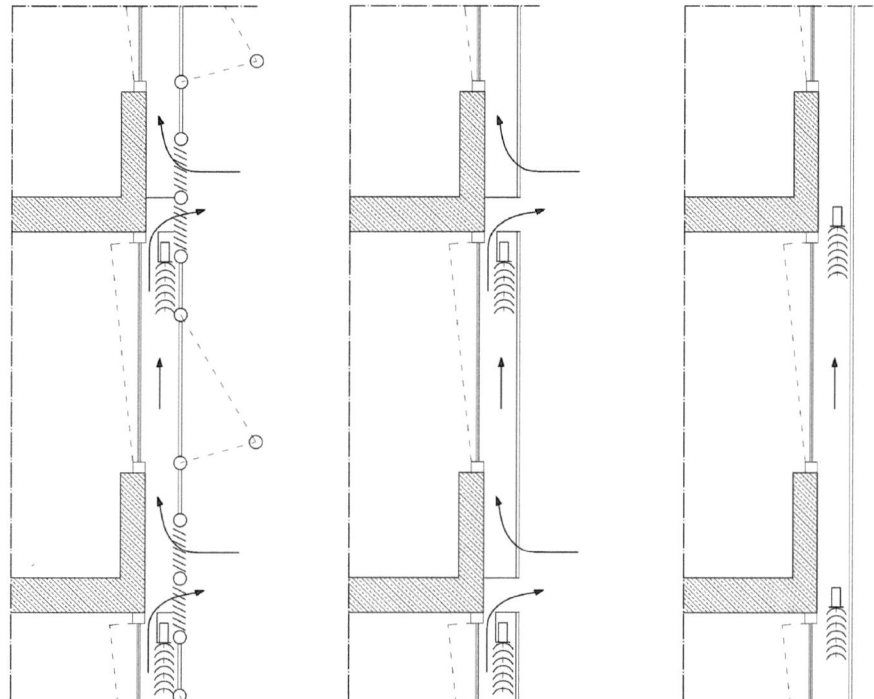

Fig. 1.7 Activation of airflows in the cavity determined by the external screen, detecting: the application of the aerodynamic devices (fixed, passively operating, or adjustable, actively operating), the ventilation of the sections in passive form without aerodynamic devices, the ventilation of the sections in passive form by the "chimney effect"

allowing the creation of a ventilated cavity or as a *buffer strip* for the thermal accumulation and insulation, reduces the wind pressure and allows the opening of the windows related to the internal curtain (even at high levels of the building, allowing the exchange of air by the natural ventilation) (Fig. 1.9).

The functioning of the multiple-skin façades systems provides that:

- during the summer season, by the day, the vertical passive ventilation conducts upwards the heat generated in the external cavity, recalling (through the opening of the window related to the external curtain) the flow of air consequent to the opening of the window in the opposite and parallel curtain;
- during the summer season, by the night, the vertical passive ventilation recalls the flow of air resulting from the opening of the window in the opposite and parallel curtain and conducts it up to the cavity, cooling the internal spaces;
- during the winter season, with the enclosure of the windows and the covering of the open horizontal grid, the thermal accumulation chamber creates (due to the "greenhouse effect") a heated air section (directed to the heating of the internal spaces) and thermal insulation (Oesterle et al. 2001).

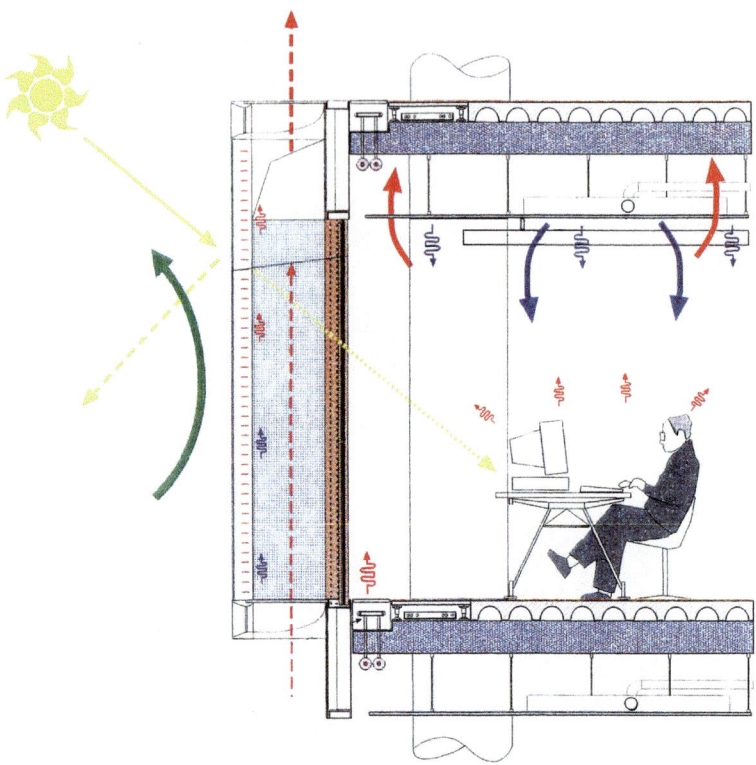

Fig. 1.8 Functioning of the multiple-skin façades system: the ventilation is guided by the regulating wing devices and by the opening-closing of the in-door frames: the combined opening generates the inflow of airflows (as convective circulation) and the natural exchange of air, supported by the upward motion in the cavity of the intermediate components; the enclosure of the air inflow elements realizes the system as a ventilated façade, with the simultaneous thermal-insulating action of the *buffer zone*. © Courtesy of Gartner

The constitution of multiple-skin façade systems is combined with the typological and constructive characters of the prefabricated façade systems, in the module type (defined as unit system). In this regard, the profile sections of the mullions and transoms are extended (by means of the longitudinal tubular cavities) up to the projection of the retaining devices of the external glazing screens. In this way, the main operation of the ventilation modes is mainly passive, considering the activation of the convective airflows for each module (provided, on the internal surface, with the opening windows) (Wasiska et al. 2014) (Fig. 1.10).

The scientific study, in its thematic framework and dissertation on the subject, intends to analyze and support the application of the multiple-skin façades systems as devices for increasing the sustainable performance of (new and pre-existing) buildings in a passive manner. The study assumes the objective of identifying and detecting, through the analysis of case studies (especially addressed through the

1.2 The Functional Development of the Multi-layer Façade Systems …

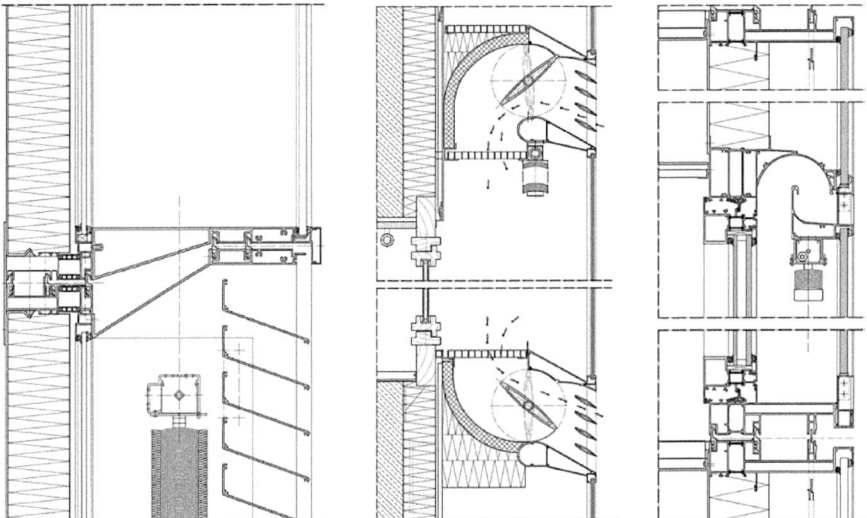

Fig. 1.9 Application of the external glazing screen and functional procedures: configuration of the flaps for the passive ventilation is passive; configuration of the flaps for the calibrated ventilation; flap configuration for the hybrid ventilation, with the lower inlet section fixed and the upper outlet section adjustable. © Courtesy of Hydro

procedures of knowledge and functional understanding by means of executive re-elaboration), the potentialities and criticalities of the applications under examination: this within a scenario that still lacks an all-encompassing scientific contribution, even if synthetic, of the main types of multiple-skin façades and the possibilities of their development.

The scientific study is developed in two main areas of research:

- the analysis of the main functional and construction modes of the multiple-skin façades, proposing their classification with respect to the executive constitution and control of the airflows within the cavities;
- the analysis of the opportunities and criticalities with respect to their diffusion also by means of non-customized components, technical elements and devices.

Therefore, the scientific study (the results of which are interpreted from the point of view of previous research), aspires to become a possible reference on the basic theoretical and constructive knowledge, in the form of a technical guide, for the analysis of the multiple-skin façades and their subsequent development through:

- the detection of a series of construction and functional methods that can be the object of diffusion, technology transfer and improvement for mass production;
- the constitution of a document in which the graphic diagrams, the executive models (also with the support of clear, selected and focused illustrations of real cases) and applications of technology transfer, system evolution and technical hybridization can constitute a reference for the interaction between expert skills in the design,

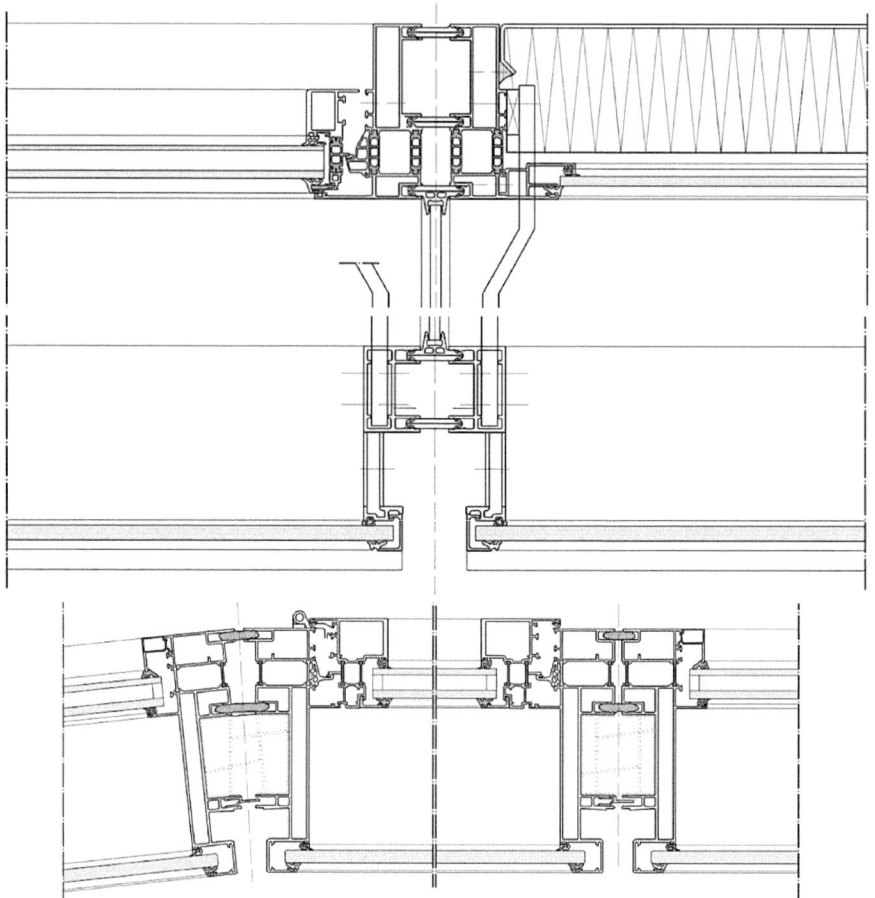

Fig. 1.10 Application of the external glazing screen: connection of the ventilated cavity achieved through the double section with the mirror joints between the mullions, which are supported externally by the brackets attached to the internal façade; mullions and transoms extended up to the projection of the devices supporting the external glazing screens. © Courtesy of Hydro

production and construction fields and expert skills in the physical field, in the modelling and simulation of the actual functional criteria: this is in order to calibrate the dimensions of the cavities, to optimize the types of transparent surfaces related to the internal façades and external shading, to proceed with the component design of the devices directed to the capture and conduction of the airflows;
- the constitution of a basic document for learning, observing and subsequently setting up the technical interfaces between the building structures (new or pre-existing) and sustainable applications of the multiple-skin façades;
- the formulation of methodological knowledge aimed at creating the cognitive and applicative basis for the theoretical and professional training of the façade

designer or façade manager (often performed by structural engineers or technical-professionals within the production and construction companies).

The development, production, and implementation of functional devices integrated with the envelope surfaces adopt methodologies and procedures associated both with the environmental and ergonomic design of interior spaces and with the energy calibration and reduction of consumption attributable to technical systems. The functional devices for multi-layer envelope systems are defined according to:

- the mediation between light transmission and thermal conduction (without compromising transparency) and the control of incident solar radiation (through the use of shading equipment or selective coatings);
- the regulation of ergonomic and energy conditions by the reflection, capture, and diffusion of external environmental loads, involving:
  - the contributions related to solar gain conditions and the reduction of thermal losses (due to the high thermal conductivity and low thermal inertia of glazed envelopes), with natural light regulation and facade overheating mitigation;
  - the development criteria aimed at selective environmental control;
  - the dynamic calibration of radiant energy fluxes;
  - the procedures for converting luminous loads through *eco-efficient* interaction.

The study materializes within the multiple-skin façades, which are defined by the acquisition and transformation of external environmental loads, based on:

- the definition of design and operative processes characterized by a high degree of flexibility in the adoption of construction techniques, components, and materials aimed at functionally adapting and establishing the overall physical-technical efficiency of the building or its constituent parts;
- the identification of criteria and tools to guide both design and execution through advanced procedural and productive techniques, marked by operative modalities that allow for specific decisions in functional adaptation interventions to meet pre-established quality standards.

The functional and compositive application refers to architectural solutions in which attention to resources and external loads becomes a formative factor of the building's shape, achieving effectiveness not only from an energy or environmental comfort perspective but also in expressive terms. The application entails connection procedures based on the use of frames and structural joint elements capable of combining the multiple functions of performance calibration, considering:

- the analysis of the construction section of vertical envelope enclosures, aimed at examining the physical and mechanical consistency of the interface sections for connecting supports or frames;
- the articulation and offering of interface solutions, with the objective of coordinating production outcomes and contemporary experimentation with the versatility and potential of their application;

- the definition of connective procedures between properties belonging to specific sections, and shape, physical, and constructive correlations.

The design development is linked to processes aimed at environmental, ergonomic, and energy definition, correlating with the functional management of interior spaces, contributions of air-conditioning systems, and consumption containment for heating, cooling, and lighting needs. The application criteria derive from the analysis of assessments related to specific environmental content, such as:

- the climatic conditions, related to latitude and the inclination of solar radiation;
- the building typologies, with reference to constructive norms linked to the climatic and environmental context, that is, in relation to the (geometric, dimensional, and quantitative) arrangement of openings in perimeter enclosures;
- the construction characters, in terms of heat storage and dispersion capacity, and the spatial and distributive configuration of interior spaces that direct airflow according to varying temperatures;
- the lighting requirements of interior spaces and according to external sky conditions, seasonal variations, and the optical properties of transparent enclosures.

The design and execution processes generally aim:

- to ensure the homogeneous diffusion and distribution of luminous contributions generated by solar radiation into interior spaces, through the use of devices capable of reflecting and transmitting fluxes toward surfaces far from the façades;
- to reduce glare conditions (particularly present when solar radiation is at a low angle during winter), whether direct or reflected from specular surfaces;
- to provide protection against thermal gains generated by solar radiation (direct, diffuse, or zenithal), to mitigate temperature increases in interior spaces;
- to reduce solar irradiance during periods of high temperature, where the use of shading devices (with a low *solar factor*, e.g., $g = 0{,}20$) involves limit (by up to two-thirds) solar gains responsible for overheating;
- to contribute to the reduction of thermal dispersion in window and façade systems during cold periods, with devices characterized by low air permeability;
- to support the generation of a ventilated cavity (in passive form), contributing to reduce internal temperatures;
- to apply enclosures that still allow for a partial transmission of light radiation, using diaphragm devices and avoiding systems that create opaque façades;
- to attenuate the impact of ultraviolet radiation, which can generate surface alteration and fading of objects inside the spaces.

References

Abrantes V, Rangel B, Amorim Faria JM (eds) (2017) The pre-fabrication of building facades. Springer, Cham

Aksamija A (2013) Sustainable façades: design methods for high-performance building envelopes. John Wiley & Sons, Hoboken, NJ

Altomonte S (2004) L'involucro architettonico come interfaccia dinamica. Alinea, Florence

Boswell K (2013) Exterior building enclosures. Design process and composition for innovative facades. Wiley & Sons, Hoboken, NJ

Daniels K (1994) The technology of ecological building. Birkhäuser, Basel-Boston-Berlin

Daniels K (2003) Advanced building systems. A technical guide for architects and engineers. Birkhäuser, Basel

Eren Ö, Erturan B (2013) Sustainable buildings with their sustainable facades. IACSIT Int J EngTech 5(6):725–730

Gunawardena T, Mendis P (2022) Prefabricated building systems. Des Construct Ency 2:70–95

Hamid AA (2012) Design and retrofit of building envelope. Bookbaby, Pennsauken, NJ

Hausladen G, de Saldanha M, Liedl P (2008) ClimateSkin. Building-skin concepts that can do more with less energy. Birkhäuser, Basel

Herzog T, Krippner R, Lang W (2004) Fassaden Atlas. Institut für Internationale Architektur, Munich

Kim G, Schaefer L, Kim JT (2012) Development of a double skin façade for sustainable renovation of old residential buildings. Indoor Built Environ 22:180–190. https://doi.org/10.1177/1420326X12469533

Knaack U, Klein T, Bilow M, Auer T (2007) Façades. Principles of construction. Birkhäuser, Basel

Lovell J (2010) Building envelopes. An Integrated Approach (Architecture Briefs). New York, Princeton Architectural Press

Mohsen A (2020) Design to manufacture of complex building envelopes. Springer, Cham

Nastri M (2021) Future façade systems. Technological culture and experimental perspectives. In: Paoletti I, Nastri M (eds) Material balance. A design equation. Springer, Cham, pp 83–103. https://doi.org/10.1007/978-3-030-54081-4_8

Oesterle E, Lieb R-D, Lutz M, Heusler W (2001) Double skin facades. Prestel, Munich

Paoletti I, Nastri M (2023) Executive design of the façade systems. Typologies and technologies of the advanced building envelopes. Springer, Cham, pp 115–137. https://doi.org/10.1007/978-3-031-44893-5_6

Schittich C (ed) (2001) Building skins. Birkhäuser, Basel-Boston-Berlin

Schmid F, Cseh X, Rohrer E, Teich M (2018) Double skin façades: Boundary conditions, challenging examples and developments. Glass Arch Struc Eng 2(5–6):103–112. https://doi.org/10.1002/cepa.914

Slessor C (1997) Eco-tech. Sustainable architecture and high technology. Thames & Hudson, London

Stazi F (2019) Advanced building envelope components. Elsevier, Amsterdam

Syed A (2012) Advanced building technologies for sustainability. Wiley & Sons, Hoboken, NJ

Wasiska I, Surjamanto W, Aswin I (2014) Natural airflow performances of double skin facade types. J Arch Built Environ 41(2):65–72. https://doi.org/10.9744/dimensi.41.2.65-72

Yonghuort L, Mohd RI (2022) Aptitudes of double skin façade toward green building within built environment. J Adv Res Fluid Mech Therm Sci 100(3):146–170. https://doi.org/10.37934/arfmts.100.3.146170

Chapter 2
The Functional and Executive Typologies of the Multi-layer Façade Systems

Abstract The study examines the constitution of the multiple-skin façades systems according to the scientific classification based on a functional, typological and constructive identification. The multiple-skin façades systems are analyzed respect to the experimental and in prototype form applications, the performances to the environmental loads and the physical, passive and active procedures of capturing the convective flows within the cavity inside the double-skins. The study focuses on the specific constitution of the sustainable methods in the form of the box-windows systems, of the shape systems, of the shaft-box systems and of the corridor systems. The configuration of the building envelope systems that function interactively and reactively with respect to the external environmental loads is articulated with respect to the procedures of capturing, transmitting (in the interior spaces) and expelling the convective airflows; at the same time, the functional configuration observes the possibilities of realizing thermal and acoustic buffers to increase the performance of the façade together with the reduction of the consumption for the climatization needs. The study focuses on the experimental and customized constitution procedures of building systems, components, elements and technical interfaces. In this respect, the scientific development is determined by the explanation of the functional devices aggregated to the façade modules (applied to the perimeter curtains or generated in the form of unit systems) that allow the interaction with external climatic stresses and comfort requirements in the interior spaces. Therefore, the analysis is expressed through an in-depth study of the construction methods, connection types and equipment that configure multi-layer façade systems and environmental devices with integrated and passive functioning.

Keywords Classification of the multiple-skin façades systems · Interactive and reactive environmental envelopes · Integrated and multi-layered façade composition · Passive technical interfaces · Executive design of functional façade equipment

2.1 The Methodology of Identification and Analytical Classification of the Multi-layer Façade Systems

The study of the sustainable methods of the multiple-skin façades systems is based on the scientific classification, on a functional, typological and constructive basis, of the main constitutive and environmental interaction modes of the external and internal loads. The scientific classification is achieved through:

- the analysis of the experimental and in prototype form applications, often realized according to customized solutions;
- the examination of the components, technical elements and interfaces between the systems and the environmental loads;
- the consequent categorization established above all by the physical, passive and active procedures of capturing the convective flows within the cavity established by the double surface of the façade and the functional devices (Gelesz and Reith 2011; Omrani et al. 2016; Bostancioglu and Onder 2019).

In particular, the study of the sustainable methods considers the implementation of the controlled ventilation according to the in-depth study of the airflows regulation devices, through the identification of:

- the opening and closing elements of the ventilation devices, which allow the regulation of the temperature and speed of the air inside the cavity;
- the transversal partitions, placed horizontally or vertically, which delimit the contiguous components that make up the multiple-skin façades systems;
- the aerodynamic control for the input, conduction and output of the airflows;
- the separation between the component units, in order to prevent the outflow of air from one unit from being reintroduced into the ventilation duct of the contiguous unit, assuring the inflow of air from the outside without the risk of it being affected by the air expelled from the interior spaces (Shameri et al. 2011; Al-awag and Wahab 2021).

The analytical research with respect to the experimental applications leads to the identification of the main functional and construction types constituted by:

- the sustainable methods in the form of the box-windows façade systems (2.2);
- the sustainable methods in the form of the multistorey façade systems (2.3);
- the sustainable methods in the form of the shaft-box façade systems (2.4);
- the sustainable methods in the form of the corridor façade systems (2.5).

2.2 The Multi-layer Functional and Executive Typology of the Box-Windows Façade Systems

The composition of the multiple-skin façade systems according to this category is specified through:

- the application of the internal façade curtain, the external glazing screen and the interposed cavity (calibrated with respect to the specific airflows ventilation);
- the location of the lower and upper openings for each module, which allow the entry of external airflows and the exit of the exhausted airflows, providing the ventilation both inside the cavity and in the internal spaces;
- the horizontal division of the cavity between the internal façade curtain and the external glass screen;
- the vertical division both between two contiguous modules and with respect to the vertical development (where the continuous divisions provide to avoid the acoustic transmission).

The process by which the natural ventilation occurs involves the air induced from outside being diverted and forced into a circular convective motion by the flow generated by the solar radiation transmitted inside. In this way:

- the part of the radiation is reflected and diverted by the sunshade foils;
- the upward rotational flow is sucked in and conveyed outside through the opening at the top of the window frame (Ding et al. 2005).

In addition, the application of the shading device inside the ventilated cavity allows the thermal load to be conveyed towards the outside, preventing it from affecting the cooling and air exchange processes of the internal spaces. The functioning of the system involves the passive vertical ventilation of the cavity between the continuous internal façade curtain and the outer glazed panel: the ventilation is achieved by capturing the airflows (due to the "chimney effect") through the open joints between the outer glass panels and at the fixing devices (Figs. 2.1 and 2.2).

The system type includes, according to the sustainable passive functioning:

- the internal façade curtain, realized by the steel sheet (acting as a vapour barrier and fire barrier element), the thermal insulation layer and the aluminium sheet cladding, together with the window with aluminium frame and double-glazing, which can be opened with a tilt and turn mechanism;
- the external screen (supported by the aluminium brackets), made from the individual glass modules with the open joints for the ventilation of the cavity.

In the case study of the double envelope system applied to the façades of the Andromeda Tower in Wien, designed by Wilhelm Holzbauer, characterized by an elliptical floor plan (for a 29-storey extension reaching the height $= 110$ m), the continuous typology is determined through:

- the internal cladding components, applied to the reinforced concrete building perimeter sections, consisting of:

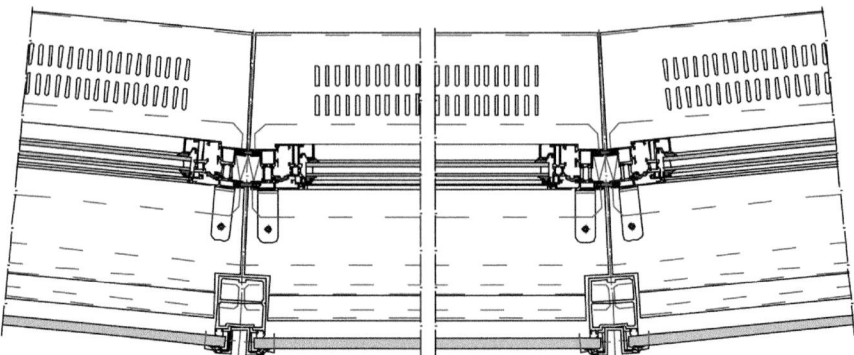

Fig. 2.1 Sustainable methods of the box-windows façade systems, where the cavities of each unit (provided with openable frames in the inner curtain) are divided by vertical partitions

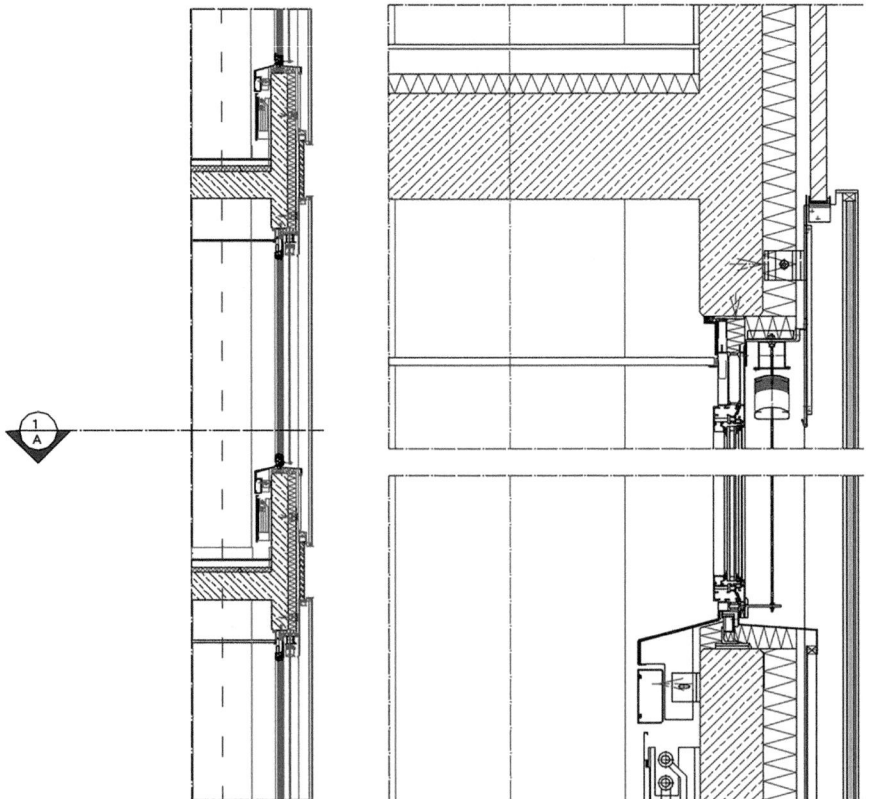

Fig. 2.2 Sustainable methods of the box-windows systems: the external glazed screen is applied at the level of the opening windows, while the module includes the thermal insulation layer and the outer cladding to enhance the laminar airflows

2.2 The Multi-layer Functional and Executive Typology …

- the internal galvanized steel sheet (thickness = 1.5 mm, acting as a vapour and fire barrier element), the mineral fiber insulation (thickness = 90 mm) and the expanded aluminium metal grid front sheet;
- the window with aluminium frame and double-glazing, which can be opened with a tilt and turn mechanism;

- the external curtain, made of single glass modules (width = 1.600 mm, variable according to the elliptical configuration, height = 3.440 mm and thickness = 10 mm) with open joints for the ventilation of the cavity. The external screen is supported by the aluminium sheet anchors (thickness = 8 mm), to which the solar shading (protected by the aluminium grid and at the top by the steel sheet, thickness = 2 mm) and the metal grid sheet are attached (Figs. 2.3 and 2.4).

The constructive processing for the vertical ventilation in the form of the box-windows façade systems involve:

- the assembly of the curtain wall system outside the main load-bearing structure (e.g. in reinforced concrete): the construction implies the mechanical connection to the steel bracket elements or adjustable flanges (by bolting to the profiles embedded in the casting, halfen type) jointed to the perimeter building sections or to the (lower and/or upper) surfaces of the floor slabs;
- the assembly between the profiles (horizontal and vertical) of the curtain wall system and the anchoring elements, followed by the joint of the window frames;
- the assembly of the profiles, horizontal and vertical (in steel or aluminium), supporting the structure in between the cavity and the subsequent assembly of the profiles or devices (in aluminium, box section, "C" or "U"-shaped) for the connection of the external glass enclosures.

In the case study of the double envelope system applied to the façades of the *Triangel Building* in Berlin, designed by Kleihues + Kleihues, the main curtain (fixed to the reinforced concrete horizontal structures by means of steel joints) is realized by hinged windows, made of aluminium profiles (Fig. 2.5).

The composition of the multiple-skin façade systems according to this category considers the application of the continuous transparent screen (as second skin) placed before the façade curtain, made of the thermal break aluminium profiles and insulation glass (as first skin). At the modular axis, the vertical steel blades cross the fire resistant insulation: these blades hold the glass of the second skin façade and divide it into independent units. The natural ventilation of the hollow space of each unit is achieved by means of the fire resistant insulation together with the shape of the front grid, which will allows the airflows. The venetian blinds do not effect the change of the air, while keeping its characteristics of sun screen (Fig. 2.6).

The type of the box-windows system with a horizontal and vertical division is also articulated through the unit façade typology, defined by prefabricated modules that are independent of each other both on a functional level (as far as the ventilation of the cavity is concerned) and on a production and construction level. The units are equipped with the air inflow and outlet openings, in which the ventilation flaps are placed to prevent incoming and outgoing air currents from mixing. The application of

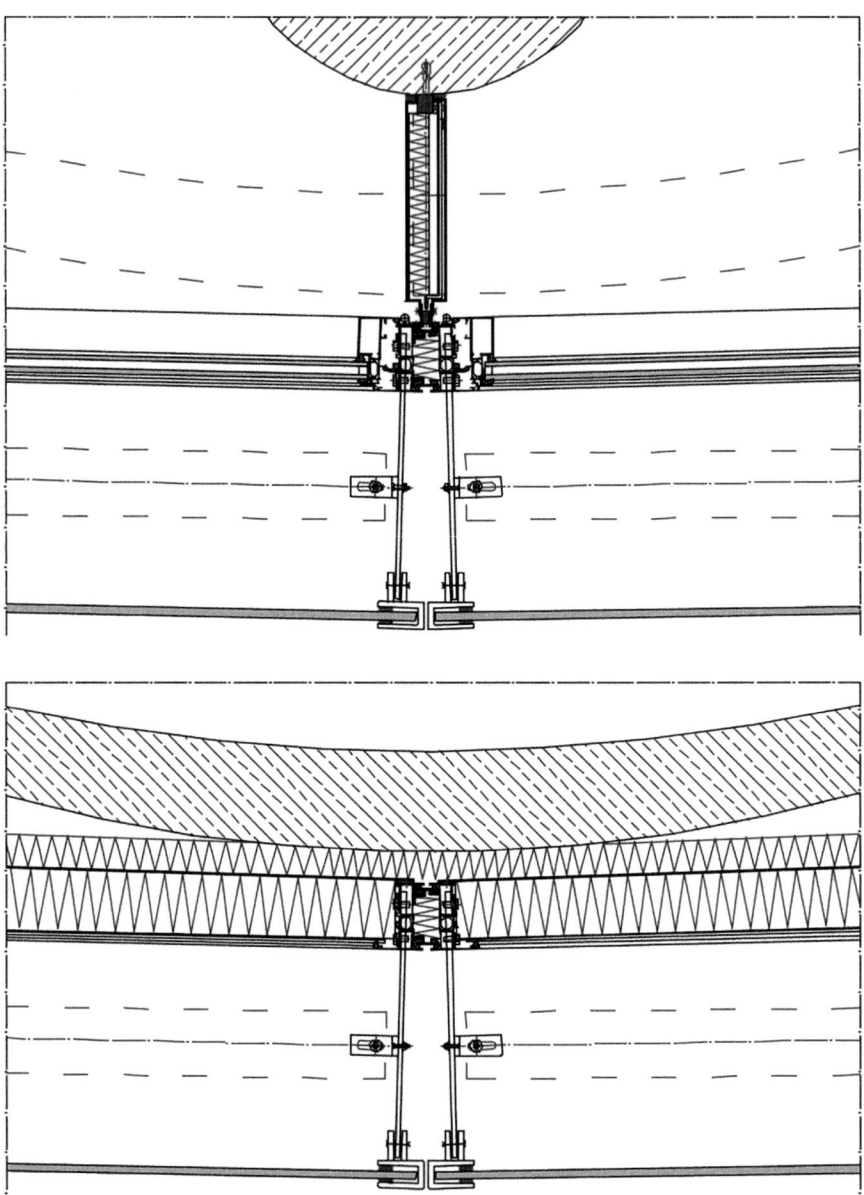

Fig. 2.3 Wilhelm Holzbauer, *Andromeda Tower*, Wien. Cavities inside the double envelope units divided by two vertical partitions and placed on the multi-layer sections

2.2 The Multi-layer Functional and Executive Typology …

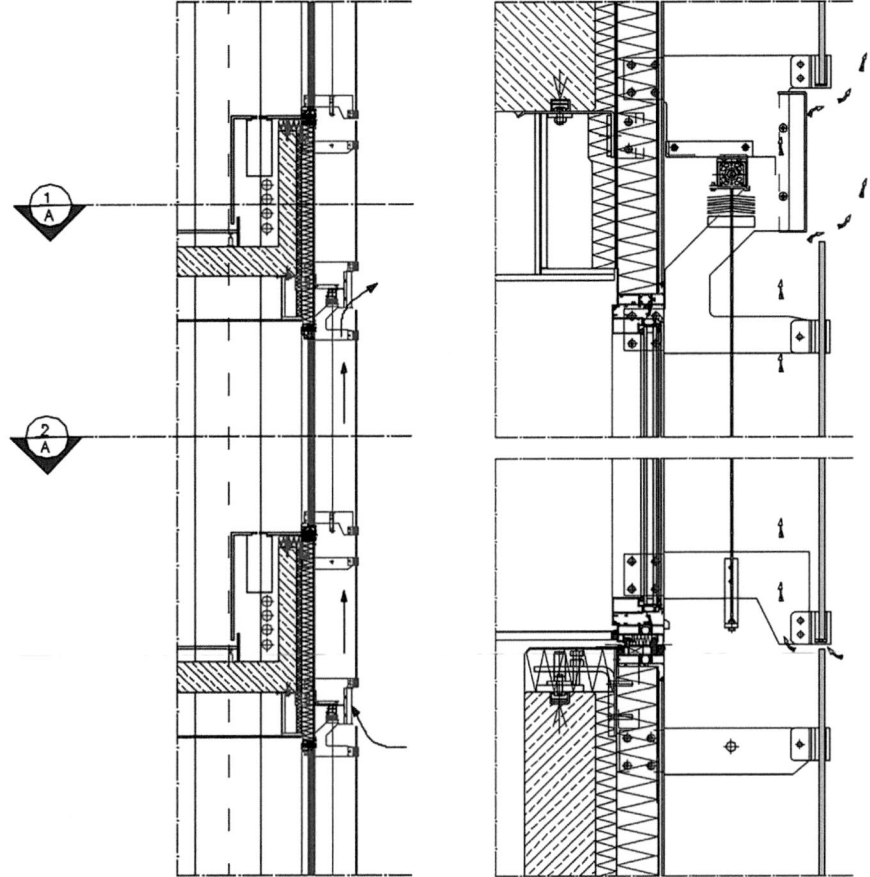

Fig. 2.4 Wilhelm Holzbauer, *Andromeda Tower*, Wien. Application of the external glazed screen at the full inter-floor level, while the module includes the thermal insulation layer and the outer cladding to enhance the laminar airflows. The internal frame in aluminium and double-glazed glass can be opened with a tilt and turn mechanism, while the external curtain is made in tempered glass (sustained by the aluminium sheet anchorage). The connection between the column and the vertical frames is made of double steel plate with plasterboard and sound-insulating mineral fiber

the unit systems is defined according to the objective of regulating thermal, lighting and internal ventilation conditions, balancing the environmental comfort in the inner spaces and controlling the energy consumption: the functioning allows the regulation of the ventilation, heat loss, solar radiation and natural lighting by the use of the sun shading in the cavity. Specifically, the systems provide, through the opening of the internal window frame, for the regulation of the air inflow and outflow, according to the aerodynamic action performed by the wing profiles:

- for the capture of the convective flows (in the lower position) from outside the façade towards the inside of the ventilated cavity;

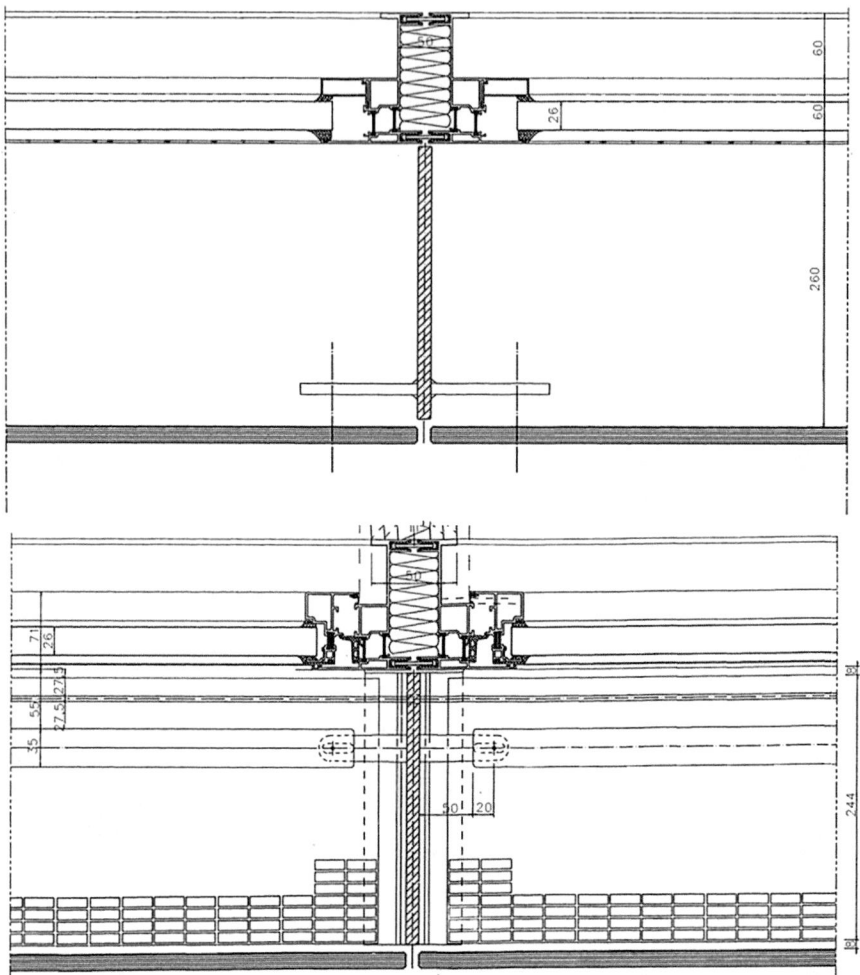

Fig. 2.5 Kleihues + Kleihues, *Triangel Building*, Berlin. Composition of the inner perimeter curtain executed by the mullions shaped in mirror form (according to the joint between the two rear and front cavities), integrated by the profiles for the attachment of the fixed or opening windows

- for the outflow (in the upper position) from inside the façade towards the outside (Khoshbakht et al. 2017).

The unit system type is often formed by the vertical ventilation components, made up of two main sections:

- the internal panel defined by the opaque section (in front of the horizontal structures) with the thermal insulation layer and the tilt-and-turn window with double-glazing enclosures;
- the external screen in toughened glass (minimum thickness = 6 mm);

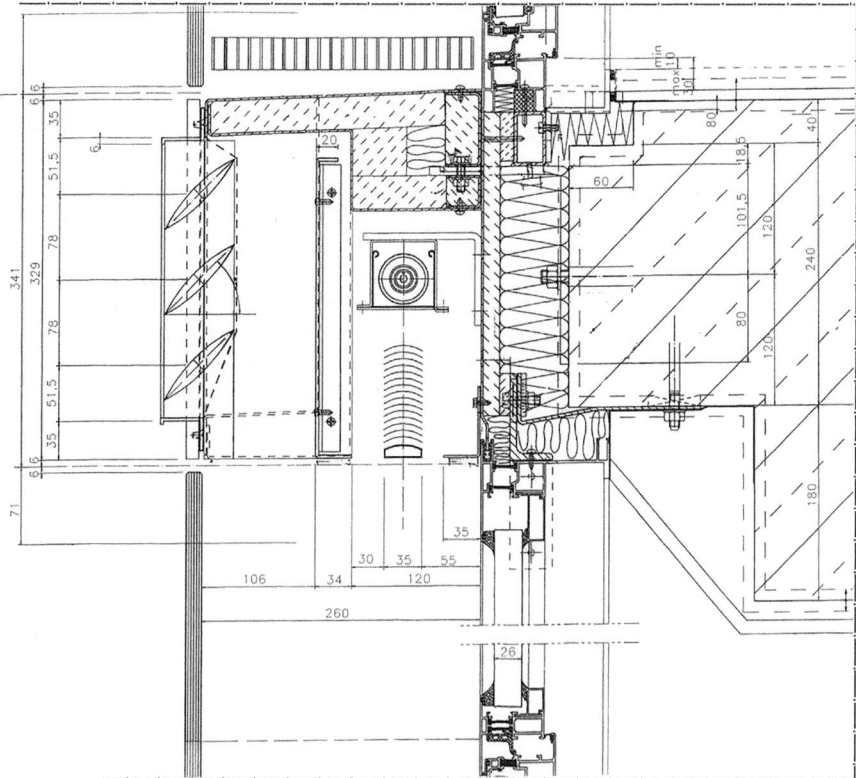

Fig. 2.6 Kleihues + Kleihues, *Triangel Building*, Berlin. Sustainable methods of the box-windows systems. The functional and construction model includes: • the extrados and intrados flanges to support and connect the windows framings; • the opening windows to the inner curtain, fire-resistant layering, external glass screen and aerodynamic flaps to guide the airflows; • the vertical dividing steel blade between the subsequent modular units, realizing the support for both the perpendicular brackets (for the fixing points connection of the external toughened glass screen, thickness = 15 mm) and the structure for the ventilation grid; • the assembly of the glass panels by means of the fixing points which are articulated to allow for expansion and free movement, involving the insertion into the perforations and locking

- the extruded aluminium wing profiles for capturing the convective airflows (in the lower position) from the outside towards the inside of the ventilated cavity and, subsequently, for the outflow (in the upper position) towards the outside.

The functioning of the system provides, through the opening of the internal frame (hinged at the bottom and opening towards the inside), the regulation of the air intake and outflow acting, in a combined manner, with the ventilated cavity. The air is drawn in from the cavity towards the interior spaces, at a level slightly above the floor level, and is expelled through the vent at the top of the façade module (Chou et al. 2009).

The assembly techniques involve the construction by means of box brackets to the upper wings of the edge beams, providing for:

- the connection of a first component, by means of a "T"-shaped profile connected to the right vertical profile, to the box bracket;
- the connection of the next component by means of the interlocking joint defined by the shapes and grooves protruding from the vertical profile of the previous component (which fulfils the function of lateral support).

The processing of the multiple-skin façades in the form of the box-windows systems also considers the type capable of achieving the vertical ventilation driven by the interstitial intake of the external convective airflows. In the case study of the double envelope system applied to the façades of the *Victoria Headquarter* in Düsseldorf, designed by Hentrich, Petschnigg and Partner, the vertical ventilation is achieved through the circular sections located in the external mullions. The unit system components are defined by two pre-assembled sections, which constitute a single functional module realized by:

- the internal façade made up of two opaque sections, a lower one (on which the aluminium frame of the tilt and turn window, with double-glazing, is grafted) and an upper one (interfacing with the external profile of the horizontal structure and with the ventilation device), contained internally by an (aluminium) layer and covered externally by a glass panel;
- the external shading composed of the single panel of toughened glass, supported by the double perforated profile mullions (in aluminium) and connected to the façade plane including, at each inter-floor level, the ventilation devices.

The functioning of the passive ventilation is by the "chimney effect" and involves the intake of the external convective flows by means of the ventilation device that introduces the air into the cavity and, through the opening of the window frames, into the interior spaces. The component is completed by the presence of the solar shading (electrically operated with a centralized control), made of aluminium with an external anti-glare coating, which allows the absorption of part of the incident solar radiation, expelled through the airflow before it can be transferred into the interior spaces by the opening of the windows. The ventilation of the cavity between the two curtains is achieved through:

- the inflow of air through the circular holes (diameter = 60 mm) placed on both profiles of the mullions (realizing an open area of 0.12 sqm for each component);
- the outflow of the air through the ventilation devices (height = 450 mm), which are characterized by two inclined deflecting wings that guide the ventilation inside the cavity towards the outside.

The assembly techniques for the construction of the systems with interstitial capture of the external convective airflows involve:

- the connection of the upper part of the components to the horizontal structure by means of the mechanical fastening to the steel angle element (connected, by

bolting, inside a slot on the upper section of the floor slab): the joint is made by means of the coupling of "T"-shaped elements which are in turn connected to the left vertical profile (in aluminium) of the upper opaque infill section;
- the application of the slots, at the perimeter of the flanges, such as to allow the joints to engage diagonally to achieve the angular opening of the mullions;
- the lateral assembly of the components, through the connections between the vertical profiles of the opaque infill sections and between the two vertical profiles of the external mullions, which integrate the insertion of the transverse glazing partition;
- the connection of the upper components (following the installation of an entire level of the façade system), by means of the joint between the horizontal aluminium profiles related to the opaque infill sections (with neoprene gaskets) and between the horizontal aluminium edge profiles;
- the closure of the slot, at the upper profile of the horizontal structures, containing the connective devices to the steel angle, by means of an aluminium flap integrated with a waterproofing band (Fig. 2.7).

2.3 The Multi-layer Functional and Executive Typology of the Multistorey Façade Systems

The composition of the multiple-skin façade systems according to this category is specified through:

- the application of the intermediate space between the inner and outer layers that is adjoined vertically and horizontally, where the cavity may extend around the entire envelope without any intermediate divisions;
- the ventilation (air-intake and extract) of the intermediate space that occurs via large openings near the ground floor and the roof;
- the possibility to close the cavity at the top and bottom, during the heating periods, to exploit the conservatory effect and optimize the solar-energy gains;
- the arrangement of the casement opening light that depends on the ventilation and cleaning concept chosen for the façade;
- the application of the external skin that is set independently in front of the inner façade, suspended by multiple types of frames and brackets, where the cavity can be ventilated in all the directions (Ahmed et al. 2016).

The multiple-skin façade systems with continuous cavity is composed according to the homogeneous and progressive development of the space interposed between the internal enclosure and the external skin, considering, with respect to the traditional ventilated wall configuration, the possibility of segmenting and articulating the façade by means of adjustable openings: this for the introduction and for the expulsion, even partial, of the convective flows contained during the upward airflow (in any case generated by the "chimney effect"), on the basis of appropriate geometric

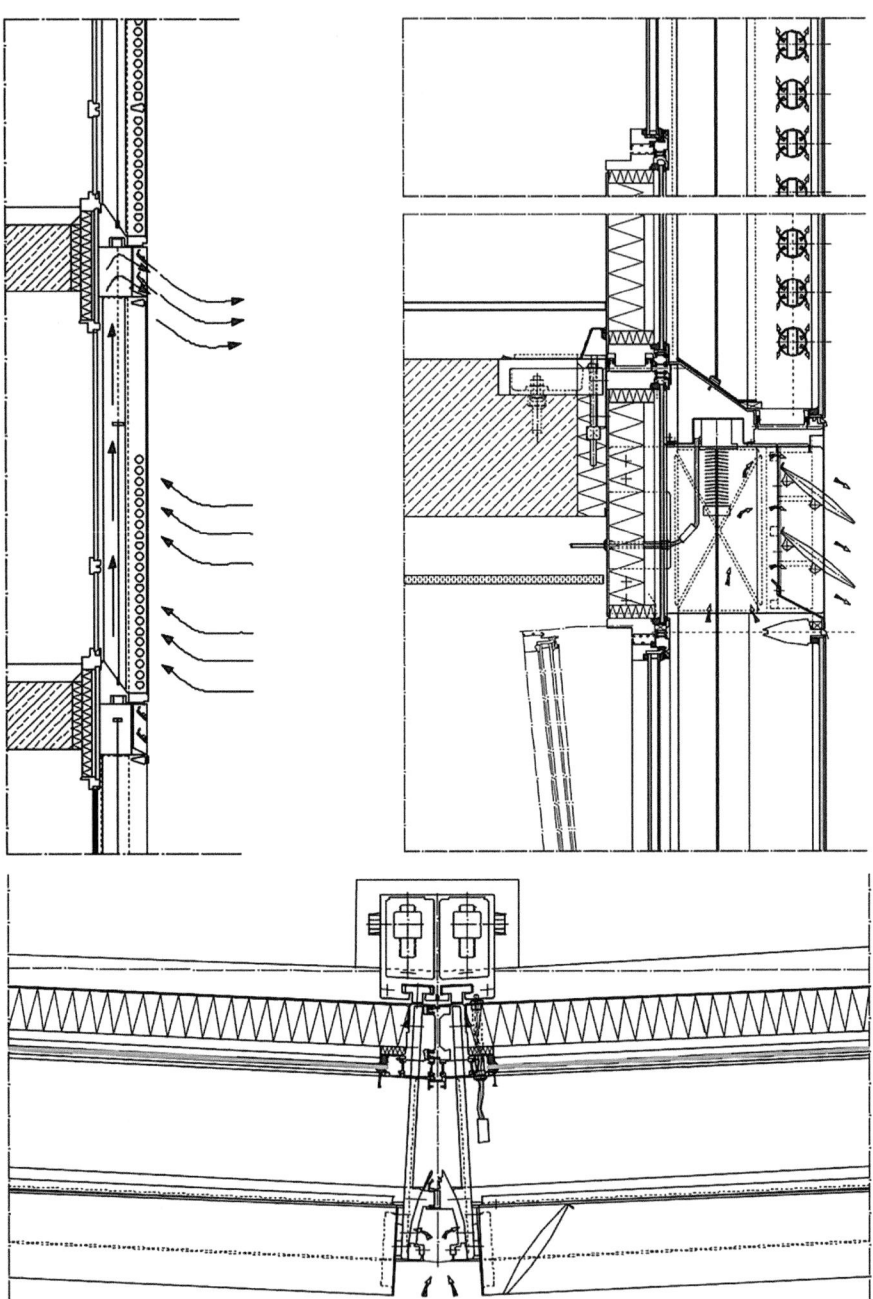

Fig. 2.7 Hentrich, Petschnigg and Partner, *Victoria Headquarter*, Düsseldorf. Sustainable methods of the box-windows systems: the ventilation of the cavity is achieved by the intake of the external convective flows through the openings in the vertical mullions, while the airflows are extracted from the cavity by the aerodynamic flaps and the cavities are divided by vertical partitions

2.3 The Multi-layer Functional and Executive Typology of the Multistorey ...

adjustments aimed at confirming the dynamic continuity and avoiding the occurrence of intermediate turbulence.

The application of the multiple-skin façades systems, in the case of the continuous cavity, is determined with respect to the development of functional and adjustable equipment by means of the activation of mechanical devices, aimed at regulating the transmission of the heat, light and natural ventilation, together with the attenuation of the external wind and acoustic loads. The system, by means of the re-radiation phenomena in the cavity (resulting from the absorbed solar radiation), realizes the homogeneous upward convective flows, which drive the ventilation of the enclosed air upwards. The composition of the multiple-skin façades according to the multistorey system, characterized by the external constitution of the adjustable glass sheets, provides that:

- during the winter period, the glass blades are closed (resulting in a sound-absorbing cavity): the solar radiation, by heating the air in the cavity, generates an insulating layer (due to the "greenhouse effect") that contributes to maintaining heat in the interior spaces and reducing the energy consumption for heating;
- during the summer period and depending on the outside temperature level, the glass blades are open, allowing the ventilation of the cavity and the night cooling of the buildings, acting in combination with the opening of the windows.

The blades of the external membrane act as a filter against the wind loads, the speed of which is high at the upper levels, allowing the windows of the internal façade to open (Safer et al. 2005) (Fig. 2.8).

The multiple-skin façade systems operate respecting the models related to:

- the winter functioning, with the aim of exploiting the heating of the air mass present in the cavity to transfer the heat to the interior spaces; that is, with the objective of directly distributing the accumulated heat, through the natural thermo-circulation established by the vertical convective flows;
- the summer functioning, with the aim of avoiding the overheating of the indoor air by removing the heat and transferring it outside.

The cavity reduces the need for heating and mechanical cooling (thickness = 700 mm). This equipment generates an upward airflow, which is also increased by the thermal gradient between the temperature in the cavity, the temperature of the incoming air and the "chimney effect" produced by the transparent curtain at the perimeter: this with respect to the objective of reducing the heat during the hot periods and controlling the energy losses, water vapour flows and frost formations on the façade plane during the cold periods. The construction of the external screen equipment contemplates the integration of the secondary frame extending beyond the internal façade: the vertical supports hold up the horizontal sections formed by the grids, which in turn form the assembly base for the mullions onto which the adjustment pivots of the glass blades engage. At the same time, the vertical supports of the outer screen are stiffened by the diagonal and horizontal profiles that react to the vertical and horizontal loads (Fig. 2.9).

Fig. 2.8 Renzo Piano Building Workshop, *Debis-C1 Building*, Potsdamer Platz sector, Berlin. Sustainable methods of the multistorey systems: behind the outer membrane of glass, which animates the reflection and refraction of the natural light with the adjustment of the blades, the terracotta cladding profiles ensure the overall architectural continuity of the elevations and visualizes the depth of the surfaces. © by the Authors

The envelope components are applied through the connection to the steel corner elements (fixed to the floor slab level), by means of the mechanical jointing devices connected to the rear curtain wall section (with the internal aluminium cladding and the external glass cladding, and interposed thermal insulating material): the enclosure and connecting profiles of the curtain wall section are equipped with the joints to both the aluminium frames of the double-glazed windows and doors and to the supports of the venetian blind sunshades. The external framing is composed of the extruded hollow plank bands made up of contiguous sections and the extruded horizontal terracotta profiles, assembled to the terracotta sheets covering the vertical aluminium mullions. The façades are made from integrated components that contribute to the comfort conditions of the interior office spaces through the adjustment of the external membranes consisting of eight overlapping sheets of glass (length = 1.300 mm, wide = 520 mm and thickness = 12 mm) (Figs. 2.10 and 2.11).

The application of the multiple-skin façades in the form of the multistorey system considers the insertion of the cavity for the purpose of insulating, filtering or absorbing the external climatic loads (above all, of a thermal nature): the envelope, in the form of an instrument of environmental interchange (of a natural, or passive type), is thus added to the perimeter enclosures to increase the capacity for controlling the internal conditions, capable of interpreting the needs of users in an *eco-efficient*

2.3 The Multi-layer Functional and Executive Typology of the Multistorey … 31

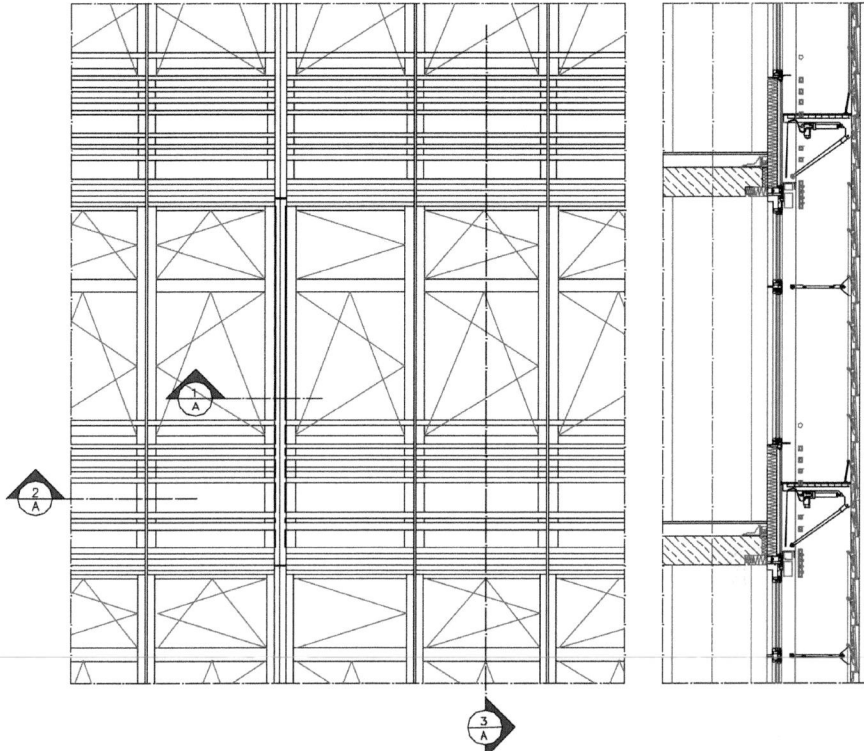

Fig. 2.9 Renzo Piano Building Workshop, *Debis-C1 Building*, Potsdamer Platz sector, Berlin. Sustainable methods of the multistorey systems: functioning of the outer blades, pivoted to the mullions (connected by the brackets to the profiles outside the façade plane), according to the adjustment guided by the electrically-operated vertical control rod

manner. Specifically, the constitution of the double-skin façade provides for the use of shading with the possibility of establishing the air changes and the thermal compensation effect of the façade. The application of the multiple-skin façade components contemplates:

- the constitution of the internal curtain wall in the unit system typology (with dimensions of 1.500 × 3.470 mm), detecting the composition of the fixed modules alternating with the modules that can be opened both for wasistas (for the climatic regulation of the internal spaces) and for maintenance requirements. The façade components are provided with the bracket elements for the connection to the rods aimed at supporting the external shading and grid;
- the positioning, contiguous to the floor perimeter sections, of the mechanically adjustable elements capable of raising (up to a height of 90°) to enhance the natural ventilation conditions of the cavity;

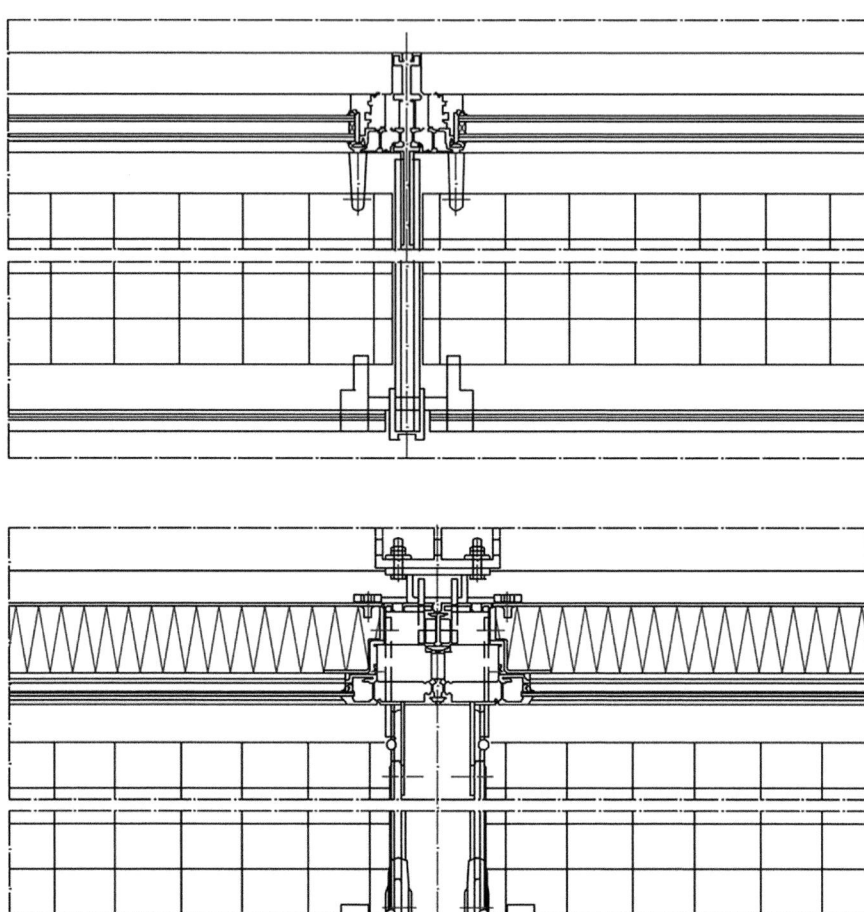

Fig. 2.10 Renzo Piano Building Workshop, *Debis-C1 Building*, Potsdamer Platz sector, Berlin. Constitution of the internal curtain wall and window frames with the double-glazing panels, of the grid in the cavity and of the external openable screen blades.: the internal curtain wall, consisting of the panels and windows (with an upper inclinable portion) with double-glazing (thickness = 38 mm, gas-filled, *U-value* = 1.1 W/m².K), integrates the hollow terracotta sheets (in front of the electrically operated sun-shading device) and the linear terracotta profiles at the stringer level

- the constitution of the external shading realized by the mechanically adjustable glazing blades (with dimensions of 1.500 × 620 mm), applied with respect to the vertical profiles. In particular, the functioning of the louvres is connected to the needs relating to both bioclimatic and fire prevention formulation, opening up to allow the passage of ventilation into the interior spaces in the event of fire (Feng et al. 2014) (Figs. 2.12 and 2.13).

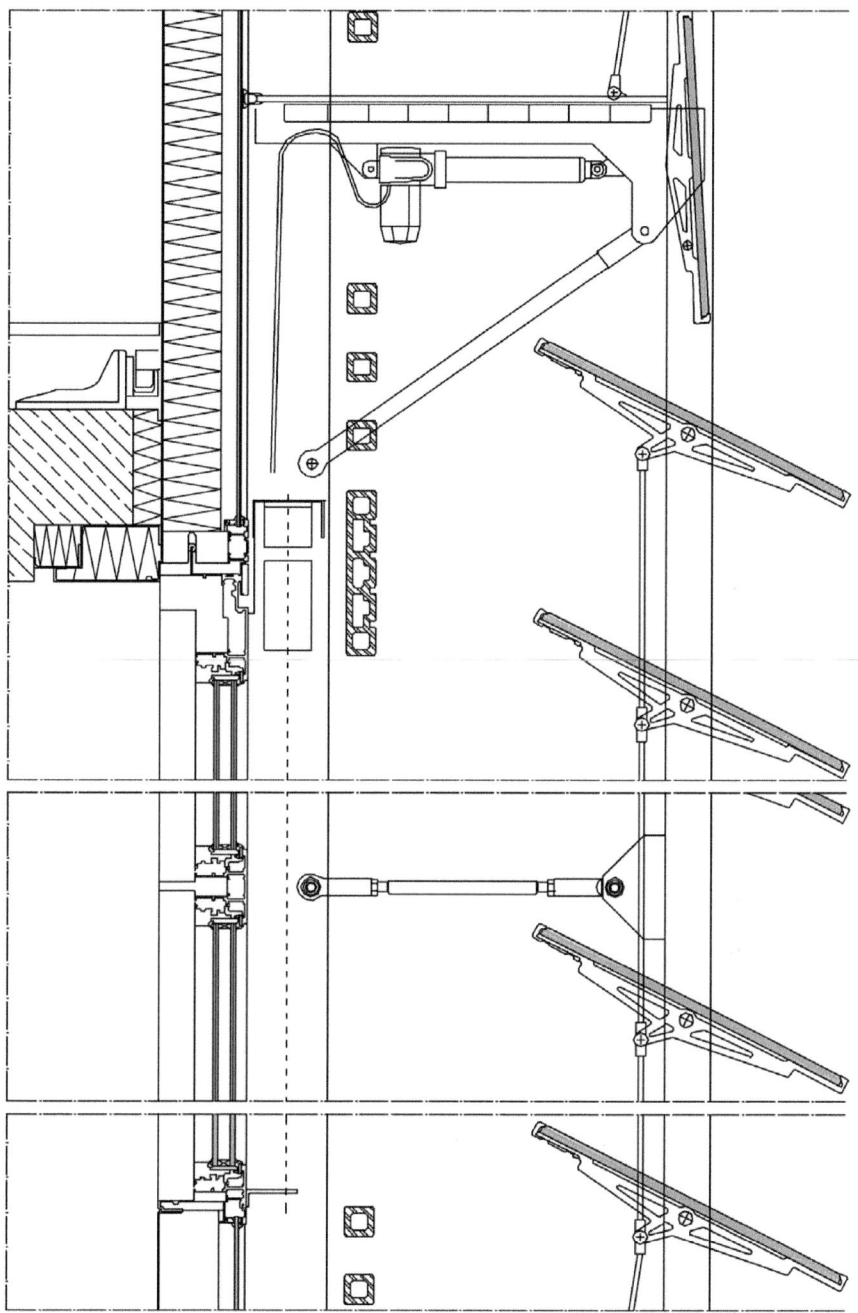

Fig. 2.11 Renzo Piano Building Workshop, *Debis-C1 Building*, Potsdamer Platz sector, Berlin. Constitution of the membranes pivoted to the mullions (connected by tubular arms to the profiles outside the façade plane) and their adjustment (by means of a hydraulic device, with an inclination of up to 70°), guided by the electrically-operated vertical control rod

The composition of the multiple-skin façade is achieved by means of the assembly of the unit system according to the mechanical application to the brackets, detecting the inter-floor form and the combined aggregation of the mullions and transoms (on the basis of the axial and specular jointing in an integrated manner to the inclusion of the linear gaskets). The profiles of the basic system are realized by:

- the longitudinal tubular section, between the two open cavities for the interface with the gaskets;
- the front chamber, for the aggregation of the couple of polyamide thermal break profiles in connection with the outer chamber;
- the outer chamber, with the ribs supporting the central sealing gasket couple;
- the glazing bead, fitted with the housing for the external grip gasket on the double-glazing enclosures and the *spandrel* panel;
- the external front section, fitted with the housing for the external gripping gasket on the double-glazing fasteners and the *spandrel* stringer panel (Hensen et al. 2002).

The composition of the multiple-skin façade is configured according to the internal façade, through the application of the unit system with respect to the assembly to the horizontal structures (by means of the galvanized steel brackets applied to the perimeter), considering the fixing hooks (or "bayonet") coupling methods of the mullion profiles. Moreover, the system provides for the aggregation to the mullion profiles of the shaped brackets for the assembly of the external frame. The construction of the external shading takes place by means of the assembly of the pre-assembled components (on the basis of the insertion of the toughened glass panels in the linear cavities included on the outside of the profiles), through the connection of the external framework. The functional coordination between the double-skin system, the main structures and the plant equipment is expressed with respect to:

- the execution of the fan coils at the perimeter towards the stringcourse sections, with the inclusion of the thermal and acoustic-insulating layers enclosed by the aluminium sheets directed to the internal vertical partitions of the transoms;
- the installation of the internal and external roller shading curtain.

The roller blind, positioned in the cavity, is operated automatically to prevent direct sunlight, maintain comfortable indoor conditions and reduce air conditioning consumption in the summer. In the absence of direct sunlight (in the morning on the west side and in the afternoon on the south side, or when the sky is cloudy), the blind retracts to allow light to enter the building, resulting in:

- an increase in visual comfort and a reduction in the consumption of energy for artificial lighting;
- during the winter season, to make use of the heat gains from solar radiation in situations that prevent increases in internal temperature and visual interference.

Fig. 2.12 Renzo Piano Building Workshop, *Intesa Sanpaolo Headquarter*, Turin. External screen equipment contemplating the space capable of including the vertical elevation structures, without interfering with the typological organization of the interiors. © Courtesy of RPBW

2.4 The Multi-layer Functional and Executive Typology of the Shaft-Box Façade Systems

The composition of the multiple-skin façades systems according to this category is specified through:

- the application of the box-window components with continuous vertical shafts that extend over a number of stories to generate a stack effect (as an alternation of the box-window components and vertical shafts segments);
- the connection of the vertical shafts, on every storey, to the adjoining box-windows by means of a bypass opening: the stack effects draws the airflows from the box-windows into the vertical shafts and from there up to the top, where it is emitted. As a means of supporting the thermal uplift, the airflows can also be sucked out mechanically via the vertical shafts;
- the require of few openings in the external skins, since it is possible to exploit the stronger thermal uplift within the stack (by contributing to the insulation against the external noise). Since the height of the stacks is limited, this type of envelope is suited to lower-rise buildings (Roth et al. 2007).

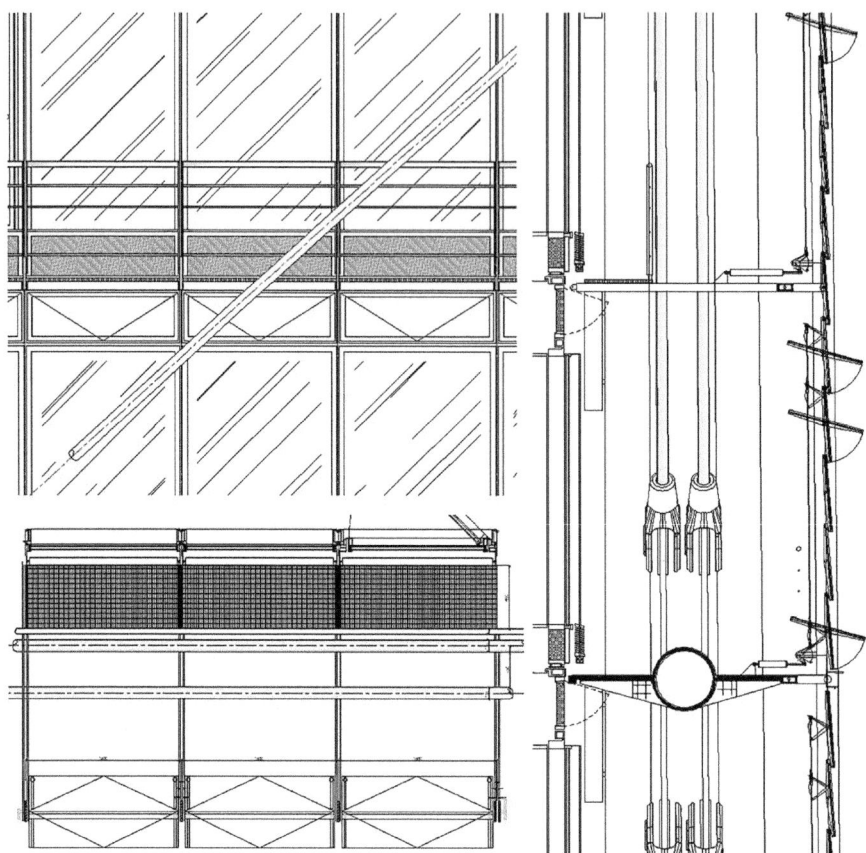

Fig. 2.13 Renzo Piano Building Workshop, *Intesa Sanpaolo Headquarter*, Turin. Sustainable methods of the multistorey systems: the technical skin provides a cavity to insulate, filter or absorb the external environmental loads, operating on the ergonomic control of the spatial conditions, air exchange and thermal compensation of the façade. © Courtesy of RPBW

The multiple-skin façades in the form of the shaft-box systems determine the horizontal-vertical division, in which the cavity is divided by vertical separating elements, which alternate the internal sections in closed wall modules and in wall modules equipped with ventilation openings. The temperature difference generated in the areas of the vertical partitions and the resulting convective airflows are used to increase the air exchange between the cavity and the interior spaces. The supply of the outside airflows takes place at the curtain modules equipped with the ventilation openings: the expulsion devices are located at the top of the side modules dividing the cavity. In this way, a depression is created that draws the exhaust air and allows the entry of outside air.

The device that encloses the modules behaves in the form of an environmentally responsive wall, capable of responding actively and organically to the climatic

2.4 The Multi-layer Functional and Executive Typology of the Shaft-Box …

loads: it operates differently during the winter period, distributing the heat accumulated by the mass of the air in the cavity, and during the summer period, with the aim of preventing the overheating in the internal spaces by removing the heat and transferring it outside.

The passive ventilation is of the vertical-diagonal type, through the integrated and functional constitution of three contiguous components, laterally enclosed by the transverse toughened glass partitions in order to conduct the air upwards and to prevent the internal flow from being reintroduced into the interspace. The double envelope units composed in this manner provide for side-to-side joining between the components and, therefore, between the ventilation devices:

- the first and third box-window components (integrated by the vertical separation panel) provide for the ventilation devices to be in the open condition towards the outside, by means of the downward rotation of the concave wing element;
- the intermediate component is open to the reception of the airflows extracted from the internal spaces (by means of the opening of the windows related to the two lateral components), by diagonal conduction and through the bypass openings arranged at the top of the vertical separation panels. It provides for the ventilation devices to be in the closed condition, by means of the upward rotation of the concave wing element, so that the inflow of the external air is prevented and the outflow of the internal air into the cavity (coming from the diagonal conduction from the two side components) is prevented.

The functioning of the double envelope system is realized through the regulation of the ventilation devices, which realize the passive ventilation for cooling or heating of the internal spaces (combined with the reduction of the heat loss by transmission): the opening and closing of these devices regulates the ventilation flow with respect to the temperatures of the air outside and inside the inner environments, to the microclimatic conditions (especially in relation to the solar radiation) and to the wind loads, affecting the energy use for air conditioning and heating of the internal spaces. In particular, the central component creates a continuous ventilated façade in which, according to the re-radiation phenomena in the cavity (resulting from the absorbed solar radiation), ascending convective airflows take place due to the "chimney effect": these drive the ventilation of the air upwards, transporting the heat generated inside and the air contained in the internal spaces to the top, where it is expelled. During the winter period, the system cavity acts as a passive heating device for the heat accumulation (by the "greenhouse effect", due to the solar radiation), improving the insulation of the curtain walls and also acting as a sound insulation barrier. The functioning observes:

- in the case of the opening of the wing devices (in the down position) and of the frames of the lateral components, the introduction of the airflows, conducted in a circle and capable of conveying the internal air upwards to the ascending convective airflows in the cavity of the intermediate components;

- in the case of closing both the wing devices (in the up position, which is also necessary for safety reasons for wind loads with a speed greater than 8 m/s) and the internal window frames, a ventilated façade is obtained.

The processing of the multiple-skin façades in the form of the shaft-box windows systems considers the passive ventilation works by means of the "chimney effect", generated by the heat radiated by the internal glass in the (inspectable) cavity between the two panels, and involves the reception of external convective flows by the air vents located at each intermediate level, on the external profile of the horizontal supporting structure. This processing is applied to the façades of the *Arag Headquarter* in Düsseldorf, designed by Foster + Partners with Rhode, Kellermann and Wawrowsky, which has a lozenge-shaped floor plan (19 storeys), with two main perspective sections, facing each other and slightly convex towards the outside, consisting of a double envelope system capable of achieving combined passive ventilation of the vertical-diagonal type within the cavity. The double envelope system consists of pre-assembled components (with dimensions of 1.500 × 3.470 mm, with the addition of upper and lower natural ventilation devices, height = 290 mm) defined by two main sections:

- the internal façade executed by the openable window frame at the floor height, with the aluminium frame and the double-glazing enclosures; the window frame related to the central layer (corresponding to the ventilated façade with continuous intake of airflows) can only be opened for maintenance operations;
- the external screen, assembled to the mullions and transoms framing of the unit, made of a single sheet of laminated glass at floor height.

The internal part is made up of aluminium frame elements on which the shapes of the horizontal profiles (supporting the internal insulating glass unit) are arranged, while the external part is made up of aluminium frame elements supporting the laminated glass panels. These elements are integrated with the vertical structural profiles, also in aluminium, consisting of a pair of internal mullions and a single external mullion to support the external glass panel and the screening device. The passive ventilation involves the uptake of the external convective airflows by means of the ventilation devices placed at the level of the horizontal structures. The elements that carry out the ventilation can be composed of:

- the two wing flaps, which guide the convective airflows towards the regulation equipment;
- the regulation equipment, with a control system and based on the rotation of the devices, which guide the closing and opening of the cavities;
- the ventilation grid (for maintenance and repair work inside the cavity).

The concave airflow inlet wing, in its open position, relates to the profile in support of the opaque rear portion of the component. The profile ends with the lower aluminium wing deflector, mirrored by the upper wing deflector, which forms the support profile for the ventilation grid. Both deflectors run between the mullions, supporting the joints that hold the external glazed enclosures. The operation observes,

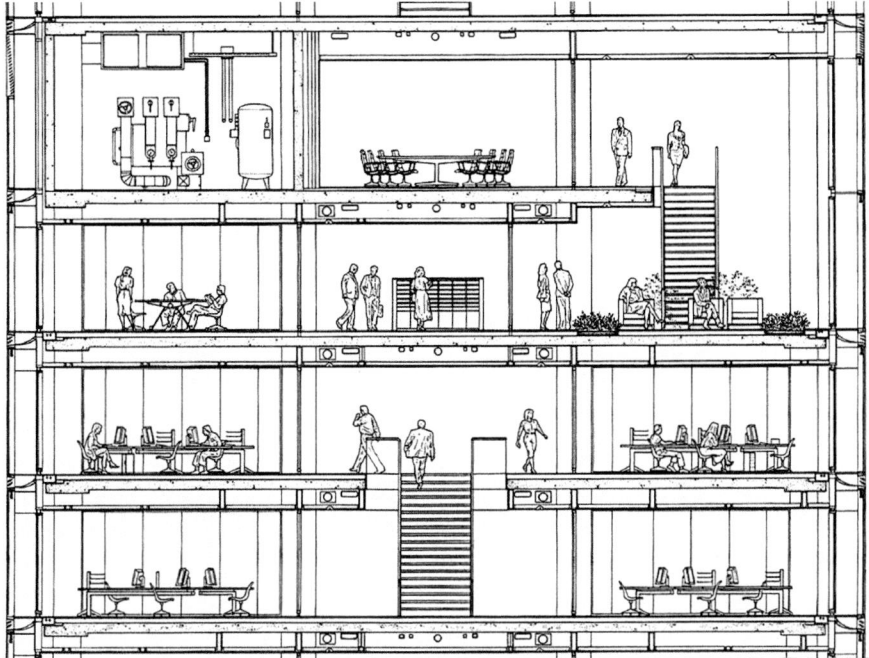

Fig. 2.14 Foster + Partners with Rhode, Kellermann and Wawrowsky, *Arag Headquarter*, Düsseldorf. Elevations enclosed by facades designed to ventilate and store heat using passive techniques, through the use of a transparent double-skin system made up of a succession of three contiguous sections. © Courtesy of Rhode, Kellermann and Wawrowsky

in the case of opening of the flaps (in the lowered position) and of the frames of the lateral components, the introduction of airflows which are conducted in a circular manner and which are capable of conveying the internal airflows to the upward convective movement in the cavity of the intermediate components. By closing both the flaps (in the raised position, for safety reasons at wind speeds above 8 m/s) and the internal frames, a ventilated façade is obtained (Figs. 2.14, 2.15 and 2.16).

2.5 The Multi-layer Functional and Executive Typology of the Corridor Façade Systems

The composition of the multiple-skin façades systems according to this category is specified through:

- the application of the intermediate space between two contiguous components closed at the level of each floor. The divisions are foreseen along the horizontal length of the corridor for acoustic, fire-protection or ventilation needs;

Fig. 2.15 Foster + Partners with Rhode, Kellermann and Wawrowsky, *Arag Headquarter*, Düsseldorf. Construction sequences. The ventilation involves the uptake of the convective airflows by means of the wing flaps characterized by a convex aerodynamic shape. The functioning observes, with the opening of the wing devices and of the frames of the lateral components, the introduction of the airflows. © Courtesy of Rhode, Kellermann and Wawrowsky

2.5 The Multi-layer Functional and Executive Typology of the Corridor …

Fig. 2.16 Foster + Partners with Rhode, Kellermann and Wawrowsky, *Arag Headquarter*, Düsseldorf. Construction sequences. The system is determined by the horizontal-vertical division, where the intermediate component is open to the reception of the airflows extracted from the internal spaces. The execution involves the lifting of the components by means of ropes attached to the profiles. © Courtesy of Rhode, Kellermann and Wawrowsky

- the application of the air-intake and extract openings in the screen placed near the floor and the ceiling: they are laid out in staggered form from bay to bay to prevent vitiated air extracted on one floor entering the space on the floor above.

In particular, the multiple-skin façades with discontinuous cavity are composed by the horizontal-type division or the horizontal-vertical division. The corridor type has the cavity segmented by the horizontal connective elements. The outside air is introduced in the lower strip of each inter-floor curtain module, while inside the exhausted air is expelled from the corridor in the upper strip. At the functional level, the ventilation flaps (also of the adjustable type) are staggered laterally or spaced vertically to prevent incoming and outgoing air currents from mixing (Baldinelli 2009). The functioning involves the diagonal passive ventilation inside the cavity through the intake of the airflows into the interior spaces (with the opening of the windows): the airflows contribute to increase in temperature (as it is heated, in the cavity, by the incident solar radiation) and to the cooling of the interior spaces, considering also that the heat absorbed by the solar shading is extracted before it can affect the temperature of the airflows entering the interior spaces.

During the winter period, the external glazed screen increases the effective thermal resistance, while during the summer period the temperature difference between the inside of the cavity and the outside produces a flow of air with a consequent decrease in the amount of the heat entering the building. In addition, the space between the two skins is used as a "greenhouse" element (to utilize the pre-heated air in the building during the winter season), put in communication with the interior through a pronounced part of the horizontal structures that act as thermal accumulators. Specifically, the system in the corridor type is composed of prefabricated single-element components divided by the vertical partitions. These floor-height components (which include the cavity and sunshade device) are applied to the external surface of the horizontal structures and are made up of two main sections. In the case study of the double envelope system applied to the façades of the RWE AG Headquarter in Essen, designed by Ingenhoven, Overdiek and Partner, the double-skin (which covers the entire external surface area, equivalent to 7.000 sqm and enveloping the 27-storey cylindrical tower building, height $= 127$ m and diameter $= 32$ m) made up of integrated type components (with dimensions of 1.970×3.590 mm) for diagonal ventilation, defined by two main sections:

- the internal panels made up of a frame with double-glazing (extra-clear Climaplus type, to avoid the greenish tint, and insulating) at floor level, supported by an aluminium frame and laterally sliding modules (manually operated). Each internal panel has one fixed and one opening module (with dimensions of 912×3.060 mm): the sliding modules can be opened (to a maximum of 135 mm) for ventilation and can be opened completely for cleaning (using a special elbow) and maintenance operations;
- the outer panels made up of an ultra-clear, reflective, tempered glass pane (with dimensions of 1.970×3.460 mm, low iron content, thickness $= 10$ mm) and connected to the supporting vertical profiles of the inner frame by means of stainless steel point joints: these joints are in turn mounted on a double pivot device

2.5 The Multi-layer Functional and Executive Typology of the Corridor …

supported by a vertical aluminium mullion (with dimensions of 50 × 120 mm) attached to the vertical profiles of the inner frame.

The double-skin system is completed by the installation of shading and anti-glare devices inside the cavity: these devices consist of aluminium foils (width = 80 mm), the centralized operation of which is calibrated to the intensity of the sun's rays for each interior space. The heat absorbed by the radiation from these devices is dissipated by the passive convective flow within the cavity (Figs. 2.17, 2.18 and 2.19).

The passive ventilation works by the stack effect, triggered by the heat radiated by the shading towards the interspace between the two panels, and involves capturing the external convective airflows through the louvers (located at the horizontal connection sections between the components) and through the aerodynamic action of the deflectors (especially, in anodized aluminium) which leads them towards the ventilation devices. These ventilation devices, with a double specular configuration, consist of wing profiles, convex downwards.

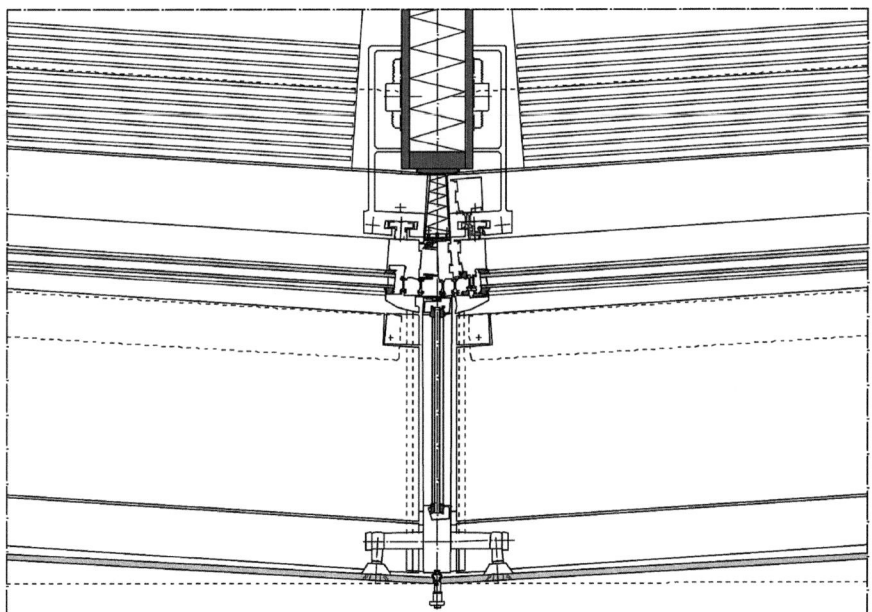

Fig. 2.17 Ingenhoven, Overdiek and Partner, *RWE AG Headquarter*, Essen. Construction of the integrated components: connection to the horizontal structure by means of mechanical fixing to the plate connected to the slab (by screwing to the horizontal profile provided in the casting); the connection is made by means of two "T" elements fixed to the section of vertical profiles relating to the internal panels; the assembly is completed by means of a steel angle profile (which rests on the slab plane and is connected to the section of vertical profiles relating to the internal panels) and a "Z"-shaped steel profile. The assembly is completed by a steel angle profile and a steel fixed to the section of vertical profiles relating to the internal cladding (supplemented by a gasket, on the external vertical surface of the slab). © Courtesy of Ingenhoven, Overdiek and Partner

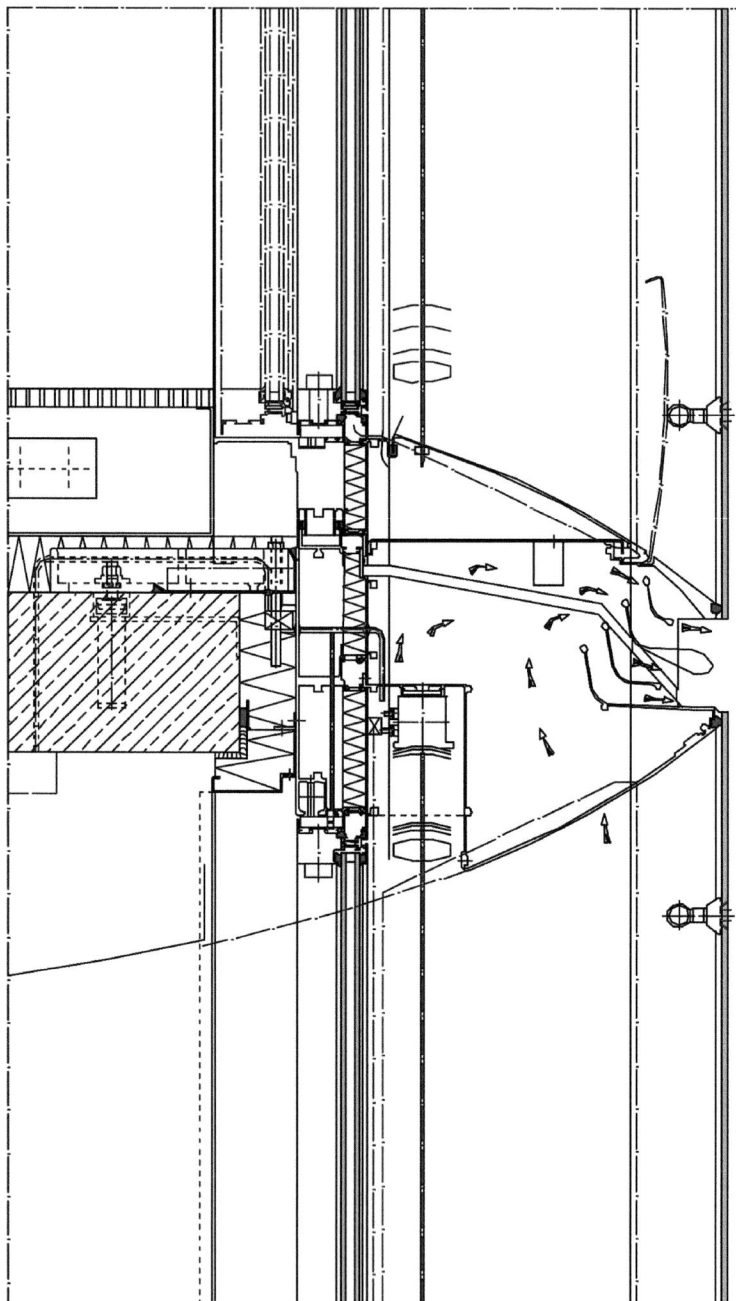

Fig. 2.18 Ingenhoven, Overdiek and Partner, *RWE AG Headquarter*, Essen. Deflectors of the integrated components, developed in the ventilation devices, are defined with a double specular wing configuration: the units provide for side-to-side coupling between two components and the ventilation flats. © Courtesy of Ingenhoven, Overdiek and Partner

2.5 The Multi-layer Functional and Executive Typology of the Corridor …   45

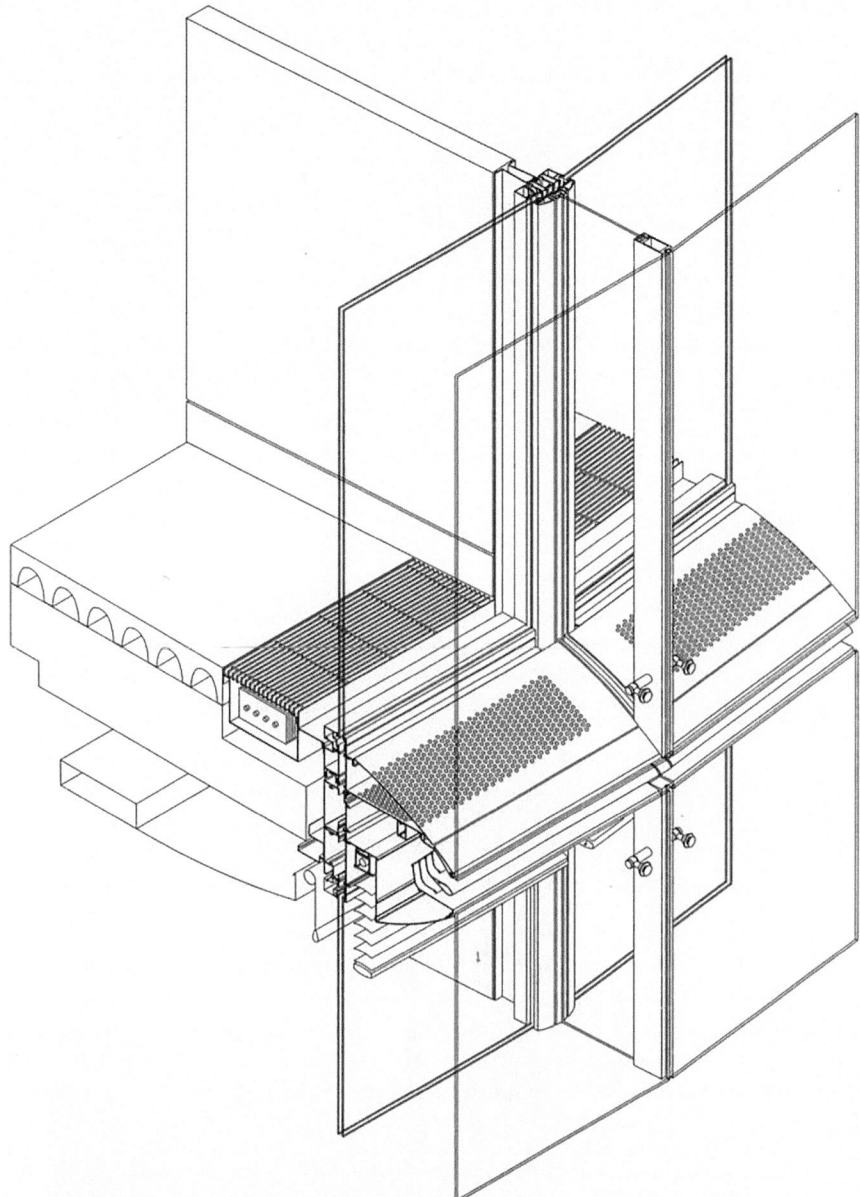

Fig. 2.19 Ingenhoven, Overdiek and Partner, *RWE AG Headquarter*, Essen. Interface section between the modules which detects the external convective airflows through the louvers. The integrated envelope system combines module frames and profiles to support passive ventilation devices: the internal façade and the external screen are joined together by the glass septum, which is connected to the perimeter mullion. © Courtesy of Ingenhoven, Overdiek and Partner

Therefore, the functioning of the system is realized through the regulation of the ventilation devices, which generates the passive ventilation for cooling or heating of the interior spaces (in this case, in combination with the reduction of the heat loss by transmission): the calibration of these devices allows the adaptation of the overall operating conditions with respect to the temperatures of the air outside and inside the built environments, to the solar radiation and wind loads, reducing the energy consumption necessary for air conditioning and heating of the interior spaces. The system consists of components defined by two main sections:

- the internal enclosure realized by a double-glazed frame at floor height, with sections that can be opened outwards;
- the external screen realized by a single tempered glass panel, connected, at each inter-floor level, to the profiles including the external ventilation devices.

The system realizes the diagonal passive ventilation through the integrated and functional constitution of two contiguous components, laterally enclosed by the transverse device of tempered glass panels in order to realize the air conduction in a diagonal manner, upwards (so as to prevent the internal airflow from being reintroduced into the ventilation circle inside the cavity). The functioning of the passive ventilation takes place by means of the "chimney effect", activated by the heat radiated by the internal glass panels in the cavity, and involves the capture of the external convective flows by means of a louver by means of the aerodynamic action of the deflectors that leads them towards the ventilation devices (Aksamija 2018). The ventilation devices of the corridor type consist of:

- a series of deflecting wings, facing outside the façade, which guide the external convective airflows towards the regulating equipment;
- the regulation apparatus, electrically operated with centralized control;
- a series of deflecting wings inside the devices, located in the lower section, necessary to direct the airflows, in a vertical direction, inside the cavity.

In the case study of the double envelope system applied to The Shard (London Bridge Tower) in London, designed by Renzo Piano Building Workshop, the multi-layer skin is determined with respect to the prospective developments established by the irregular polygonal planimetric configuration and the asymmetrical intersections between the enclosure surfaces (which covers the entire external surface area, equivalent to 55.000 sqm and enveloping the 87-storey building, height = 310 m). The integrated component (with general dimensions of 1.500 × 3.800 mm) involves:

- the construction of the unit components, provided with the openable window frame, the internal glass curtain (installed with the aid of thermal break aluminium framing) and the external glass screen, with the shading inside the cavity. This device, which limits heat gain from solar radiation, contributes to reducing the effects of the external curtain wall temperature on the internal rooms;
- the assembly to horizontal structures of metal corrugated sheet with concrete casting, using the halfen profiles for aggregating the brackets (to which the aluminium frames are applied using rear "bayonet" coupling);

2.5 The Multi-layer Functional and Executive Typology of the Corridor …

- the thermal break aluminium frame consisting of double box-section profiles that support the internal double-glazed curtain (comprising the internal panel of extra-clear laminated glass, thickness = 44.2 mm, the cavity, thickness = 16 mm, as well as the external panel of extra-clear float glass with a low-emissivity coating, thickness = 8 mm); the assembly is secured by inserts of thermal break polyamide bars. The window, as internal floor-to-floor height, is openable in a casement style to allow access to the cavity for cleaning and maintenance operations;
- the horizontal frame at the top of the components (where the attachment interface to the brackets is filled with fireproof material), supporting the box-shaped partitions with the extension towards the external aluminium frame. This frame is used to provide structural adhesion to the extra-clear laminated glass screen (with a reflective coating of 24%, thickness = 44.2 mm). The frame is then connected to the internal supports of the unit through a series (for the two vertical sides of each component) of "T"-shaped aluminium joints to the mullions;
- the shading device inserted into the cavity, realized by the fiber glass roller blind (white-grey chrome-plated, with a 5–10% colour variation depending on the module's position along the façade), which is operated by the building management system (reducing solar radiation by 85%).

The multi-layer system allows only 0.12% of solar radiation to pass through, reducing the need for cooling systems) while sound insulation reaches a performance level of Rw > 43 dB). The double-skin façade involves the use of laminated glass external screening to optimize the functions enabled by ventilation in the cavity. The passive functioning is governed by the natural ventilation process inside the cavity (thickness = 200 mm), based on the horizontal opening between the consecutive vertical sections of the external screens.

The corridor type is able to control air exchange and the thermal compensation effect of the façade through horizontal ventilation slots. The opening slot, in particular, is defined by the plane extrados surface of the upper transom and the wing element in shaped sheet metal, inside the cavity, connected to the external frame profile related to the lower transom of the main frame. The use of double-skin reduces heat loss from interior spaces by decreasing the speed of air flow in contact with the outer panel, thereby generating a high level of thermal insulation.

The air exchange process, which is integrated with the limited dimensions of the ventilation "chimney" located between the two façade enclosures, ensures a constant flow of air with the outside environment through the slot left between the vertical sequence of screens. The system works by generating a thermally insulated section during periods of low temperature (acting as passive buffer zone), keeping the shading (located in the cavity) wrapped up in order to absorb solar gains: moreover, ventilation from outside acts against possible condensation on glass surfaces.

During periods of high temperatures, shading is carried out along the façade module in order to reduce the incidence of solar radiation. The air contained in the cavity is set in motion upwards by the absorption of solar radiation by the glass panels, the shading devices and the metal connecting elements: the flow transports and evacuates the accumulated heat to the outside (for an amount equal to 25% of the heat resulting from direct radiation in the cavity), while the ventilation acquired from the outside contributes to limiting the overheating (Figs. 2.20, 2.21 and 2.22).

In the case study of the double envelope system applied to the Unipol Headquarter in Bologne, designed by Open Project, the multi-layer skin, developed with respect to the trapezoidal plan (height = 125 m), involves:

Fig. 2.20 Renzo Piano Building Workshop, *The Shard*, London. Unit system components: thermal break aluminium frame consisting of double box profile sections. © Courtesy of Giulia Pastore

2.5 The Multi-layer Functional and Executive Typology of the Corridor ...

Fig. 2.21 Renzo Piano Building Workshop, *The Shard*, London. Double-skin functioning: reduction of radiant heat transfer and natural ventilation of the cavity (according to the horizontal opening between consecutive vertical sections of the external screens). © Courtesy of Giulia Pastore

Fig. 2.22 Renzo Piano Building Workshop, *The Shard*, London. Technical interfaces between the horizontal stiffening beam system (as outrigger beams), which reduces lateral deformation and wind action, and the double-skin component framing. © Courtesy of Giulia Pastore

- the construction of the unit components (south and east façades), provided with a discontinuous horizontal corridor cavity (width = 800 mm). The functioning of the system is divided into two phases:
  - the operation during the winter period, when the windows are mainly closed to reduce heat loss and to create a "greenhouse effect", with advantage of radiation to generate the increase of a buffer zone containing a mass of air at a higher temperature: this results in an improvement in the average radiant temperature of the façade (in the case of double-glazing) and, consequently, an improvement in the temperature of the internal spaces, as well as a reduction in heat transmission from the inside to the outside;

2.5 The Multi-layer Functional and Executive Typology of the Corridor …

- the operation during the summer period, providing for the (natural or forced) inflow and rise of convective air movements inside the cavity, with the aim of partial heat dissipation (assuming in the design contents a summer temperature 3 °C higher than the outside temperature, for a total of 37 °C);
- the design of the unit components (north side), consisting of the aluminium thermal break frame supporting the double-glazing, made up of external solar control and internal low emissivity panels, with the cavity filled with argon gas and characterized by a total thermal transmittance value of $U_w \leq 1.00$ W/sqm.K;
- the implementation of the ventilated façade (west façade), made up of the cladding surface (to cope with direct solar radiation during the summer period), with glass panels combined with photovoltaic modules and aluminium panels, and with ventilation grilles directed towards the technical compartments (Fig. 2.23).

The façade section provides the location of the automatically adjustable opening windows and doors, which interact with the arrangement of the deflectors (applied to the false ceiling) that, when opened, reduce the possible generation of undesirable airflows. The type of façade with horizontal corridors makes it possible to maintain the fire compartmentation between the different floors and allows smoke to escape through the openings in the external façade. The external shading is made up of two fixed modules of laminated glass with solar control, including the part of the corridor structured according to the orientation of the louvres for the introduction of convective airflows. The fixing points are used for connecting the passing brackets in the vertical rebate between the mullions, projecting outwards to support (by screwing) the horizontal tubular steel elements in the form of tie-rods (Fig. 2.24).

The fixing points support the steel bars from which the point-fixing devices to the laminated glass panels of the external screen (connected in linear rebate by the silicone gasket) branch off. The intrados interfaces provide the connection to the louvre blades (with tubular steel frame, single or coupled section, with flat insect screen fixing), which are directed towards the airflows in the intermediate space: in this respect, the unit system connects the aluminium profiles that support the opaque and openable transom sections, of layered constitution (after assembly by screwing through the external sheet metal cladding) (Figs. 2.25 and 2.26).

Specifically, the construction provides the connecting interfaces between the type of enclosure (in the unit components) and the sections, both outside the roof, through the brackets (anchored to the halfen profiles) from which the telescopic coupling to the mullions is exposed, and inside the roof, through the corner elements (still connected to the halfen profiles). The structural extension, which reaches the perimeter steel truss, is thermally insulated on all sides up to the rebate towards the façade: the layered transom section can be opened outwards (by mechanical operation) to allow the air to flow out of the interior spaces, with the contribution of the ceiling. In addition, within the cavity of the double-skin components, there is the integrated installation of solar shading of the blind type (with slightly convex blades), which ensures:

Fig. 2.23 Open Project, *Unipol Headquarter*, Bologne. Typological and performance development of the double-skin unit façade components (south and east elevations), the unit curtain wall components (north elevation) and the opaque ventilated façade (west elevation), with external glass cladding, photovoltaic modules and aluminium panels. © Courtesy of Schüco

- the effectiveness against direct and reflected solar radiation, with the possibility of redirecting incident solar rays towards the internal ceiling surfaces, increasing the diffusion conditions of natural light;
- the heat supply depending on the external temperature and solar radiation, being particularly effective when the external temperature is lower than the temperature of the internal spaces and when the total radiation values are low;
- the conditions of easy construction and management (through automatic regulation), since the devices are protected from wind loads and external conditions.

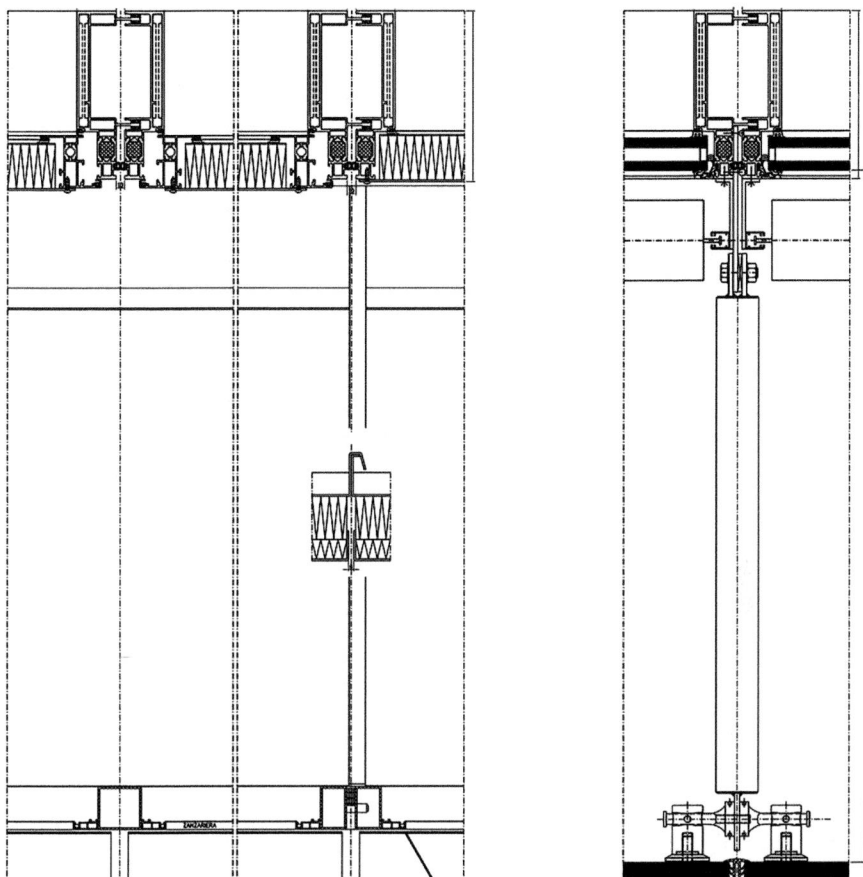

Fig. 2.24 Open Project, *Unipol Headquarter*, Bologne. Executive design of the technical interfaces between the aluminium frames (mullions, transoms) (of the cellular façade type) and the supporting elements for both the external shading and the deflector sections. © Courtesy of Schüco

2.6 The Scientific and Executive Perspectives: Potential and Criticality of Multi-layer Façade Systems Development and Application

The study, defined on the basis of scientific knowledge and the analysis of the composition of multi-layer façade systems, is characterized by the proposal of methods for their transformation and application. On this basis, the research is proposed as a tool both for in-depth typological and functional studies of integrated building envelope systems and for the identification of technology transfer procedures. The scientific study supports the efficiency or criticality in the development of the systems, demonstrating the possibilities of use. Moreover, the study of the multi-layer façade

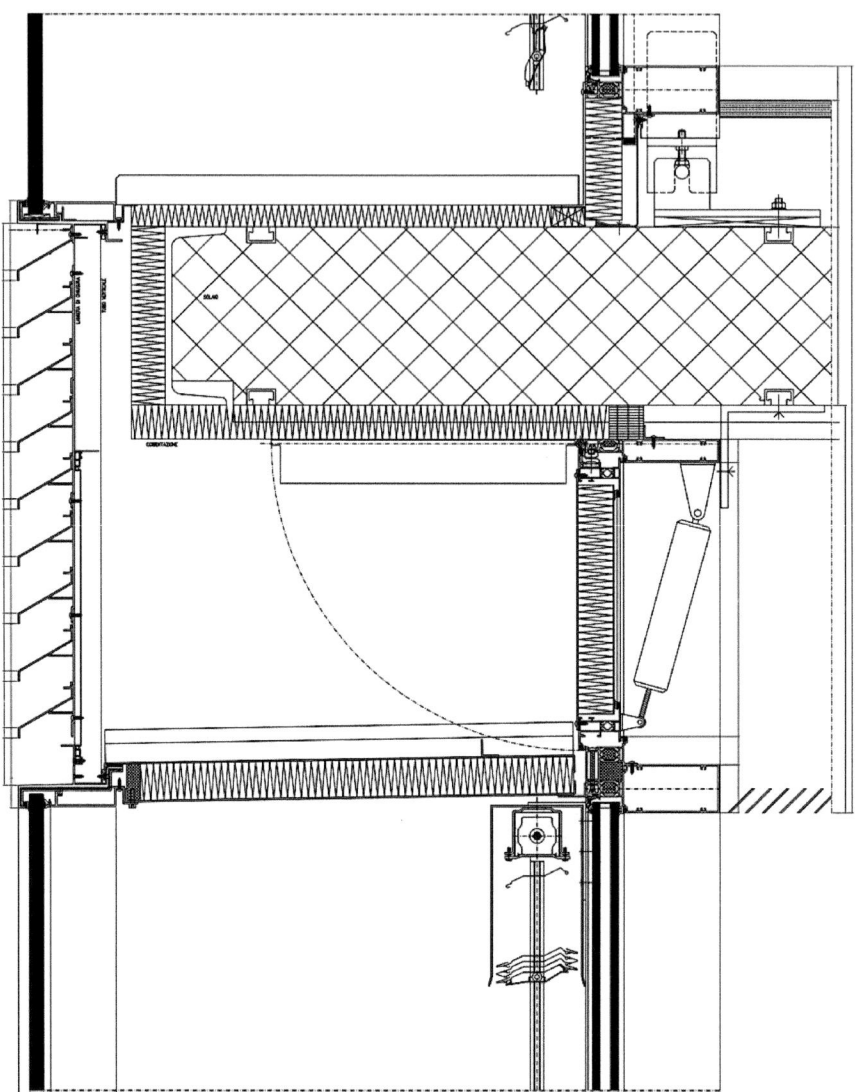

Fig. 2.25 Open Project, *Unipol Headquarter*, Bologne. Executive design of the interface details of the horizontal structural section in relation to the connections to the curtain wall, the opening fanlight enclosure and the enclosure oriented towards the external screen. © Courtesy of Schüco

systems, with respect to the scientific analysis concerning both the functional constitution (1.1, 1.2) and the typological and construction classification (2.1), on the basis of the specific methodology of identification (to the experimental and prototype applications, of the components, technical elements and interfaces towards the environmental loads), is determined through:

Fig. 2.26 Open Project, *Unipol Headquarter*, Bologne. Double envelope system construction: internal curtain with aluminium framing, interposed cavity and outer screen supported by mechanical fixing joints. © Courtesy of Schüco

- the detection of the sustainable methods directed to the physical and passive procedures of natural ventilation, light radiation calibration and thermal storage;
- the detection of the specific methods of activation, transformation and acquisition of the environmental loads;
- the detection of the opportunities and criticalities of replication, diffusion and technology transfer to the buildings of common character of new construction or subject to technological requalification (Musa and Alibaba 2017).

The use of the executive re-elaboration of the systems as a technical and research tool for modelling and functional reproduction, constructive understanding and sustainable methods allows the identification of potential developments of the multi-layer façade systems (Paoletti and Nastri 2023). The research results, supported by the identification and explanation of several case studies, consider:

- the development of the potential of sustainable methods in the form of the box-windows systems (2.2), through:
  - the constitution of systems and components of reduced geometries, dimensions and functionality, understood as system evolution and in the form of technical hybridization with respect to the reference typologies;
  - the constitution of systems and components aimed at the technological requalification of the pre-existing buildings (with frame structure and vertical development), through the replacement of the vertical enclosures with interactive systems with respect to the external and internal environmental loads;
  - the constitution of systems and components through the use of standard, mass and non-customized products, in the form of technology transfer;
- the development of the potential of sustainable methods in the form of the multistorey (2.3) and of the corridor systems (2.5), through:
  - the constitution of systems and components of calibrated geometries, dimensions and functionalities with respect to the vertical and horizontal elevation structures of new and pre-existing buildings;
  - the detection of multiple applications, for specific or complete envelope extensions to the activation of upward convective airflows by the "chimney effect";
  - the possibility of establishing calibrated applications with respect to the functions at thermo-hygrometric, lighting and acoustic levels in combined or specific form, even with respect to the façade sections with openings;
  - the possibility of applying the screen components in the form of modules or in the form of prefabricated unit systems;
  - the adaptability of the shading solutions either by means of customized or standard and non-customized types of framing, or by means of different layering of glass surfaces or other cladding components;
- the criticality related to the development of the sustainable methods in the form of the shaft-box systems (2.4), as it determines:

- the need for a vertical extension such as to allow the combined aggregation of two types of components with different functioning and according to three types of execution (i.e., the series of central modules for the convection of the airflows related to the exhaust air and the side modules both for the acquisition of the external airflows and for the conduction towards the central modules);
- the need for complex design and production of the devices for acquiring external airflows, which require specific aerodynamic calibration against the environmental conditions of the context (Høseggen et al. 2008).

References

Ahmed MMS, Abel-Rahman AK, Ali AHH, Suzuki M (2016) Double skin façade: the state of art on building energy efficiency. J Clean Energy Tech 4:84–89

Aksamija A (2018) Sustainable façades: design methods for high-performance building envelopes. J Façade Des Eng 6:1–39. https://doi.org/10.7480/jfde.2018.1.1527

Al-awag EAN, Wahab IA (2021) Perspectives in double-skin façade (DSF) advantages and disadvantages. In Proceedings of IOP Conference Series: Earth and Environmental Science, 6th International Conference on Civil and Environmental Engineering for Sustainability (IConCEES), pp. 1–9. https://doi.org/10.1088/1755-1315/1022/1/012003

Baldinelli G (2009) Double skin façades for warm climate regions: analysis of a solution with an integrated movable shading system. Build Environ 44:1107–1118. https://doi.org/10.1016/j.buildenv.2008.08.005

Bostancioglu E, Onder NP (2019) Applying analytic hierarchy process to the evaluation of double skin façades. Archit Eng Des Manag 5(1):66–82. https://doi.org/10.1080/17452007.2018.1515062

Chou SK, Chua KJ, Ho JC (2009) A study on the effects of double skin façades on the energy management in buildings. Energy Convers Manag 50:2275–2281. https://doi.org/10.1016/j.enconman.2009.05.003

Ding W, Hasemi Y, Yamada T (2005) Natural ventilation performance of a double-skin façade with a solar chimney. Energ Buildings 37:411–418

Feng X, Yang H, Feng XY, Jin FY, Xia GQ (2014) A review of research development of ventilated double-skin façade. Appl Mech Mater:587–589. https://doi.org/10.4028/www.scientific.net/AMM.587-589.709

Gelesz A, Reith A (2011) Classification and re-evaluation of double-skin facades. Int Rev App Sci Eng 2(2):129–136. https://doi.org/10.1556/irase.2.2011.2.9

Hensen J, Bartak M, Drkal F (2002) Modeling and simulation of a double-skin façade system. ASHRAE Trans 108:1251–1259

Høseggen R, Wachenfeldt BJ, Hanssen SO (2008) Building simulation as an assisting tool in decision making: Case study: with or without a double-skin façade? Energ Buildings 40:821–827. https://doi.org/10.1016/j.enbuild.2007.05.015

Khoshbakht M, Gou Z, Dupre K, Altan H (2017) Thermal environments of an office building with double skin façade. J Greenbuilding 12(3):3–22. https://doi.org/10.3992/1943-4618.12.3.3

Musa BT, Alibaba HZ (2017) Evaluating the use of double skin facade systems for sustainable development. J Med App Biosci 9:55–72

Omrani H, Ghaffarianhoseini A, Ghaffarianhoseini A, Raahemifar K, Tookey J (2016) Application of passive wall systems for improving the energy efficiency in buildings. Renew Sust Energ Rev 62:1252–1269

Paoletti I, Nastri M (2023) Construction of the façade systems. Production and assembly procedures of the advanced building envelopes. Springer, Cham. https://doi.org/10.1007/978-3-031-49608-0

Roth K, Lawrence T, Brodrick J (2007) Double-skin facades. ASHRAE J 49

Safer N, Woloszyn M, Roux JJ, Kuznik F (2005) Modeling of the double-skin facades for building energy simulations: radiative and convective heat transfer. Build Simul 2:1067–1074

Shameri MA, Alghoul MA, Sopian K, Zain MFM, Elayeb O (2011) Perspectives of double skin façades systems in buildings and energy saving. Renew Sus Energy Rev 15:1468–1475. https://doi.org/10.1016/j.rser.2010.10.016

Chapter 3
The Technical Hybridization of the Multi-layer Façade Systems

Abstract The study examines the criteria for the integrated, combined and interactive development of multi-layered façade systems, consisting of the projection of performance and environmental modes beyond the main perimeter sections. The analysis is based on the possibilities of composition of different types of screens and external devices capable of increasing both the thermal, hygrometric, lighting and acoustic functions, as well as the perceptive and expressive contents. In this context, technical hybridization processes aim at the definition of façade modules with adjustable or fixed applications, based on the design and construction possibilities of frameworks and supports extending outside the façade curtains: these are directed at supporting multiple levels of layering and execution, calibrated according to climatic and environmental loads, through specific surface treatments. The hybridization techniques are configured through the work of aggregation against prefabricated unit system components, by which the efficiency of the work of mechanical arrangement, the definition of technical interfaces and the specific resolution of connection and operation processes is determined. On this basis, the study considers the architectural, performance and construction outcomes resulting from the technical hybridization of envelope systems, focusing on the typologies of profiles, connectors and material adaptability of external surfaces (in terms of results achievable through both customized solutions and mass-produced elements). The analysis also aims to demonstrate the potential for technology transfer from experimental implementation to application in conventional construction: the executive and constructive in-depth study, using functional and structural models, aims to enhance the development of hybrid façade types and the criteria for design and production processing, as well as the options for interaction with the needs of interior spaces and external environmental visualization.

Keywords Technical hybridization · Integrated envelope systems · Interactive and environmental operating devices · Executive design of prefabricated envelope components · Technical and functional interfaces of unit system components

3.1 The Executive and Functional Constitution of the Integrated Façade Systems

The evolution of the contemporary experimental research around the multiple-skin façade systems (e.g., in the box-window type) is determined according to the thick surface envelope, often applied for the technological requalification and energy transformation of existing buildings for tertiary and commercial uses. The envelope is defined in the form of a double-skin façade, characterized by the cavity that allows both natural ventilation and the external shading against the solar radiation and the acoustic loads. The executive and functional constitution of the external devices integrated to the façade systems is expressed through the experimental application inside the sector "Mac567" of the *Maciachini Center* in Milan, designed by Matthias Sauerbruch and Louisa Hutton. The integrated building envelope is a tool for the morpho-typological and perceptive redefinition of places subject to redevelopment and urban modification interventions. The design poetics works with the polychromy applied to the façades to produce the visual intensity of the perimeter curtains, and thus both the lightness of the reverberation induced by the cladding material, and the dissolution of the external unity through the subdivision and multiplication of coloured patterns with constant rhythms. The effect of the building is then softened by the coloured enclosures, which fragment the reading of the surfaces and accentuate the vibration on the façade planes: the chromatic spectrum is therefore very differentiated, projecting itself towards the functional and plastic configuration and towards the sensitive execution of the urban experience. In particular, the geometric fantasy defined by the mosaic of the façade modules, repeated in bright polychromy, is intended to provoke surprising perceptive phenomena, transforming the vertical enclosures of the pure parallelepipeds with pictorial afflatus.

The intervention, with the aim of connecting and integrating this space, once isolated and physically separated from the city, into the urban metamorphosis of the built context, assumes the will to create a harmonious interpenetration between the settlement and the park: this in order to make explicit the paradigms of environmental and dynamic fusion of places, according to the volumetric and, above all, perspective expression of the façade systems. The dimensions of the intervention and the envelope, interwoven within a kind of "giant" architectural order, are expressed through alignments and proportions derived from a careful observation of the surroundings. Even transparency, within a language that reduces all aspects of the façade components, is applied according to a symbiotic relationship with the greenery and light that cladds and permeates the building forms. Within the overall intervention, the sector called "Mac567" is divided into three buildings: two of these organisms are equipped with a transparent connecting structure at the level of each floor. The envelope is generated to enhance verticality through the serial combination of linear façade modules (with dimensions of 4.000×1.500 mm), in turn constituted by six sub-components unified in their geometries and interface sections. The functional and environmental design is based on the objective of bringing the spaces

Fig. 3.1 Matthias Sauerbruch and Louisa Hutton, "Mac567" sector, *Maciachini Center*, Milan. Composition and relational expression of the facades. The polychromy of the envelope curtains is conceived and applied as a perceptive means of urban redevelopment, allowing the visual brightness and the dissolution according to the series of painted modules. © Courtesy of AluK

into climatic and sensorial equilibrium with the external context (Figs. 3.1, 3.2, 3.3 and 3.4).

The chromatism on the rigorous, linear and planar surfaces is discreet and not intended as pure decoration; it is based on an original mixture of elegance, boldness and precision of the sign, brought into interaction with natural light and the contemporary perspectives of ecologically conscious architecture. This approach is supported by the need to consider the current composition and urban contexts in the form of screens and interactive, palpitating surfaces. The architectural design process involves the integration of the morphological, functional and executive interface of the glazed elements with the main structure, as well as the implementation of exact and unconstrained spatial cuts that extend beyond the defined volumes. This approach combines poetic expression with principles of sustainable construction and the exploitation of climatic loads (with the objective of reducing energy consumption and calibrating the ergonomic comfort of the workspaces) and incorporates a multi-layer façade system.

The envelope system is engineered to include two glass panels divided by a cavity (formed by structural aluminium profiles with thermal breaks) to facilitate convective ventilation. The constitution contemplates the use of unit systems components, alternating glazed panels with fixed horizontal continuous parapets, which are equipped with acoustic insulation material. The windows are protected by adjustable sunshades that automatically adapt to the sunlight's angle (Figs. 3.5, 3.6 and 3.7).

Fig. 3.2 Matthias Sauerbruch and Louisa Hutton, "Mac567" sector, *Maciachini Center*, Milan. Composition and relational expression of the facades. The painted enclosures accentuate the vibration on the façade levels according to a varied colour spectrum, creating multiple perceptual and visual phenomena. © Courtesy of AluK

The panels and sunshades are made of the same type of glass, treated with coloured opaque screen printing, which allows a high level of transparency (even when the sunshades are closed) and a high level of solar protection. In particular, the brise soleil is electrically adjustable to cut the sun's rays and deflect their direct incidence. The system enclosures consist of the following elements:

- for the internal curtain, by double-glazing panels (with an air gap, thickness = 18 mm, of which the internal panel is of the laminated type, with thickness = 6 + 6 mm, equipped with PVB separating foil), tempered HST and transparent, low-emissivity type (with *thermal transmittance value* of $U_w = 1.40$ W/sqm.K);
- for the external sunshades, by silk-screened single-glazed panels (thickness = 12 mm), rotating around the vertical axis and movable in groups to adjust the shading function. In order to obtain a system capable of combining the requirements of calibrating natural light incidence and transparency from the built environment, the silk-screening is applied to the inner face of the blades by means of a "dotting" of reduced geometries above the painted surface (Figs. 3.8, 3.9 and 3.10).

The multi-layer façade typology in unitized system form is applied directly to the load-bearing reinforced concrete structures, through the assembly to the mechanical devices projected beyond the perimeter beams. The objective of the area's redevelopment is rezlised in the vertical enclosures, which employ silk-screen printed glass in various colours and electrically openable sunshades to express the synthesis between architectural language and performance purposes (Fig. 3.11).

3.1 The Executive and Functional Constitution of the Integrated Façade … 63

Fig. 3.3 Matthias Sauerbruch and Louisa Hutton, "Mac567" sector, *Maciachini Center*, Milan. Urban settlement and projections of the envelope. The insertion expresses the characters of environmental fusion, fruition and dynamic perception of places. © Courtesy of AluK

Fig. 3.4 Matthias Sauerbruch and Louisa Hutton, "Mac567" sector, *Maciachini Center*, Milan. Urban settlement and projections of the envelope. The perimeter curtains are associated with the urban landscape through the expression of alignments, cuts and perspectives of the connecting volumes, creating a contrast with the homogeneity of the site. © Courtesy of AluK

Fig. 3.5 Matthias Sauerbruch and Louisa Hutton, "Mac567" sector, *Maciachini Center*, Milan. Geometric and relational modulation of the envelope. The polychromic components on the planar surfaces of the façades combine with the climatic, perceptive and sensorial balance of the interior spaces in relation to the external context, using research around visual effects and the "sense of three-dimensionality". © Courtesy of AluK

Fig. 3.6 Matthias Sauerbruch and Louisa Hutton, "Mac567" sector, *Maciachini Center*, Milan. Geometric and relational modulation of the envelope. The planar, polychromatic and interactive surfaces provide the executive interface with the load-bearing vertical reinforced concrete frame structure. © Courtesy of AluK

3.1 The Executive and Functional Constitution of the Integrated Façade ... 65

Fig. 3.7 Matthias Sauerbruch and Louisa Hutton, "Mac567" sector, *Maciachini Center*, Milan. Composition of the envelope system. The curtain is constituted by multi-layer façade components according to the unit system typology with two glass panels that include the naturally ventilated cavity (protected by sunshades that can be oriented to solar incidence). © Courtesy of AluK

The system is based on the combined constitution of the aluminium mullion profiles, with the box section facing outwards: this is associated (by means of the pair of compartments, also of box configuration) with the double polyamide bars stretched towards the "C"-shaped profile connecting to the frames (fixed and opening) to support the double-glazing panels; the containment of the glazing beads is performed both by the projection of the double bars from the frame profiles and by the extension of the aggregate sheets to the box section of the mullions (Fig. 3.12).

Beyond the load-bearing metal structure, the unit system components are equipped with cladding elements in the form of sunshade panels at the height of the storey, in the form of adjustable screen-printed glass panels (according to the position of the pivot pin on the horizontal centre line) or in the form of fixed panels. At the same time, the façade section beyond the horizontal structures is made with the *spandrel* enclosure with the double acoustic-insulating layer (Fig. 3.13).

The unit system components are assembled by means of fixing hooks (or "bayonet") coupling methods of the mullion profiles to the vertical pins projecting from the brackets (assembled to the slabs by means of the halfen profiles). The same mullions perform the support for the brackets projecting towards the frames (of aluminium composite type) characterized (in their vertical section):

- in the extrados section of the sill plate, by a box configuration with a planar upper surface intended for connection to the aluminium sheet and for facilitating the

Fig. 3.8 Matthias Sauerbruch and Louisa Hutton, "Mac567" sector, *Maciachini Center*, Milan. Composition of the shading devices. The double-skin façade, manufactured using prefabricated components in the unit system typology, is the solution suitable for large-scale building projects, characterized by executive rationalization procedures and performance calibration towards both internal spatial ergonomics and the reduction of energy consumption. The cladding devices in brise soleil form, at full height and extending beyond the façade components, consist of adjustable or fixed glass panels, produced with the same screen printing procedure. © Courtesy of AluK

adjustment joint of the rotatable plates. Additionally, a frontal surface is present for the attachment of the fixed screen-printed glass panels;
- in the intrados section of the sill plate, by a box-shaped configuration in the form of a transom. The "U"-shape projection faces both the adjustment joint of the rolling panels and the support of the horizontal sheet enclosure (Fig. 3.14 and 3.15).

3.2 The Executive and Functional Constitution of the Combined Performance Façade Systems

The application of the principles of technical hybridization to multi-layer façade systems is expressed in the determination of the type of envelope, defined in the form of an "augmented window". Specifically, the façade consists of the double

3.2 The Executive and Functional Constitution of the Combined … 67

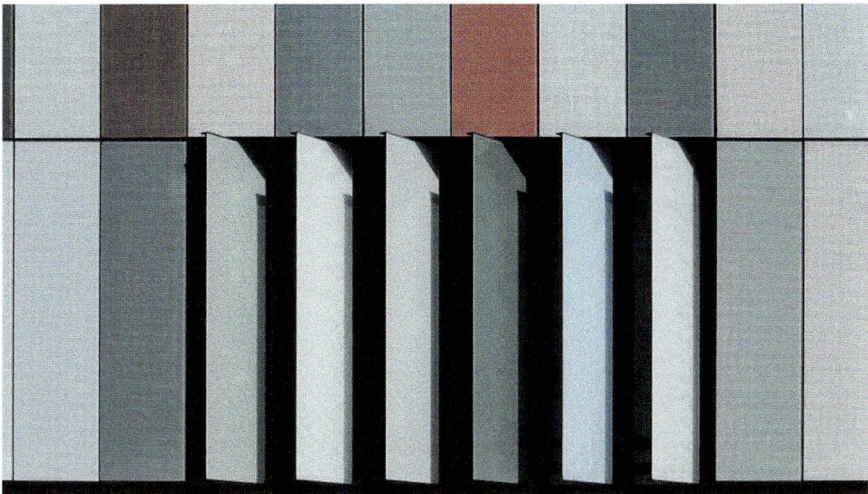

Fig. 3.9 Matthias Sauerbruch and Louisa Hutton, "Mac567" sector, *Maciachini Center*, Milan. Composition of the shading devices. The brise soleil devices are electrically adjustable to affect the sun's rays and to allow diffuse light transmission into the interior spaces, through mosaic modulation on the façade planes. © Courtesy of AluK

Fig. 3.10 Matthias Sauerbruch and Louisa Hutton, "Mac567" sector, *Maciachini Center*, Milan. Building envelope enclosure system. The construction of the external sunshades, in the adjustable type for the attachment of the rotation pin on the perimeter frame, provides for the single glass panes: these are screen-printed and movable to optimize the shading function. © Courtesy of AluK

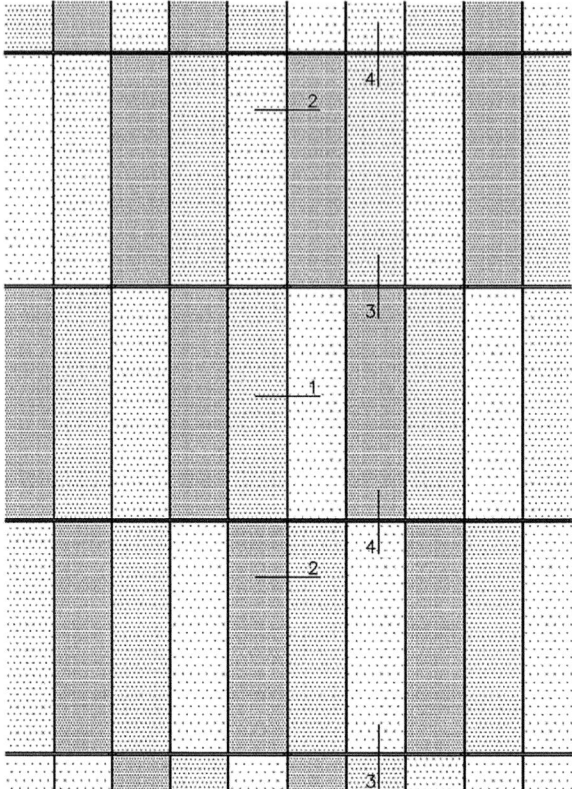
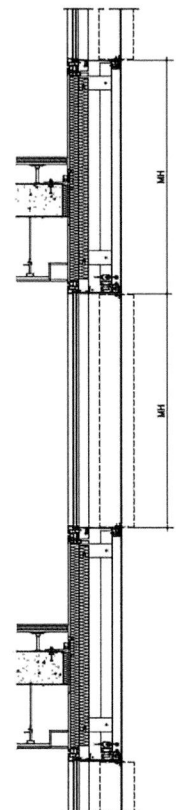
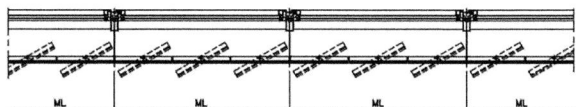

Fig. 3.11 Matthias Sauerbruch and Louisa Hutton, "Mac567" sector, *Maciachini Center*, Milan. Functional and executive composition of the multi-layer façade according to the unit system typology, integrated with the adjustable external screens. © Courtesy of AluK

envelope type, characterized by the ventilated interposed cavity that allows both the natural ventilation of the façade and the external screening against solar radiation and acoustic load. The passive functioning of the system (by means of the chimney effect and the convective airflows) is combined with an internal mechanized ventilation system, thereby reducing the energy consumption of the architectural organism in comparison with a traditional continuous enclosure section (especially during

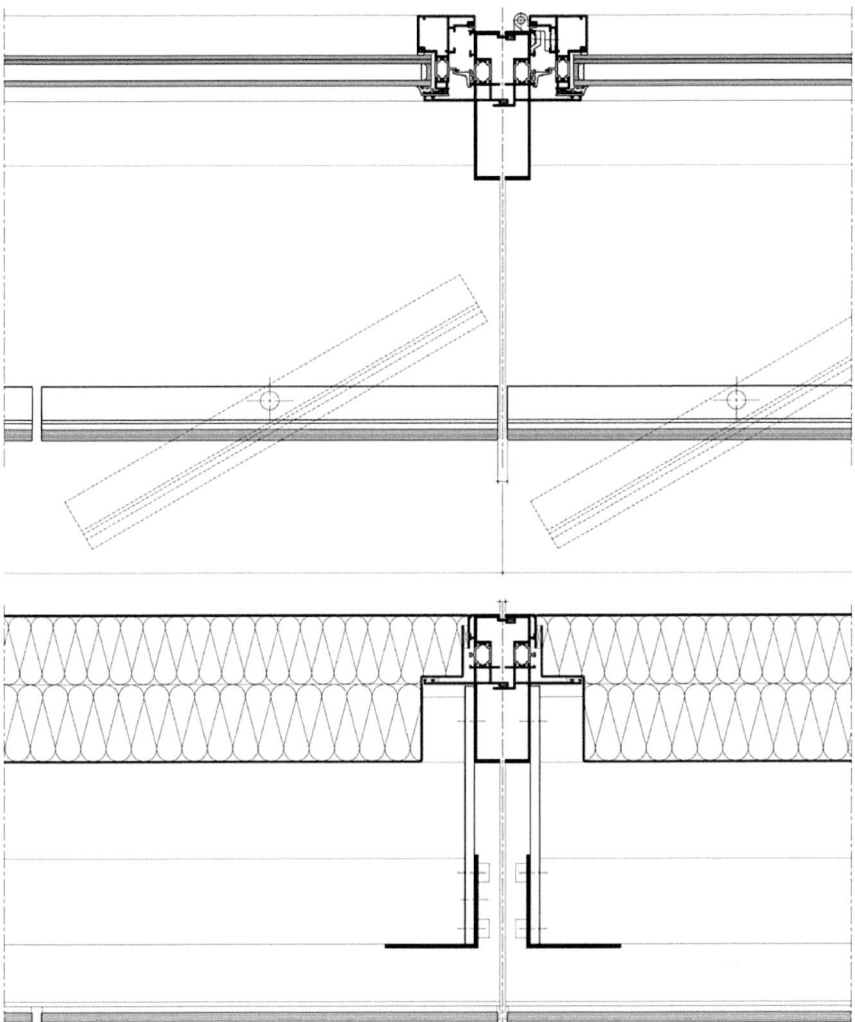

Fig. 3.12 Matthias Sauerbruch and Louisa Hutton, "Mac567" sector, *Maciachini Center*, Milan. Functional and executive composition of the façade multi-layer according to the application of the screens in both an adjustable configuration, on the outside of the opening windows, and a fixed configuration, on the outside of the *spandrel* sections. © Courtesy of AluK

summer and winter): the implementation of the "augmented window" enables the removal of the former air-conditioning system located along the perimeter, in addition to the ducts inserted into the false ceiling. The heating and cooling procedures are thereby entrusted to the series of radiant ceiling panels.

The diaphragmatic and polymorphic configuration of the "augmented window" is examined with respect to the *Société Privée de Gérance (SPG) Headquarters* in Geneva, designed by Giovanni Vaccarini Architetti. The building is characterized

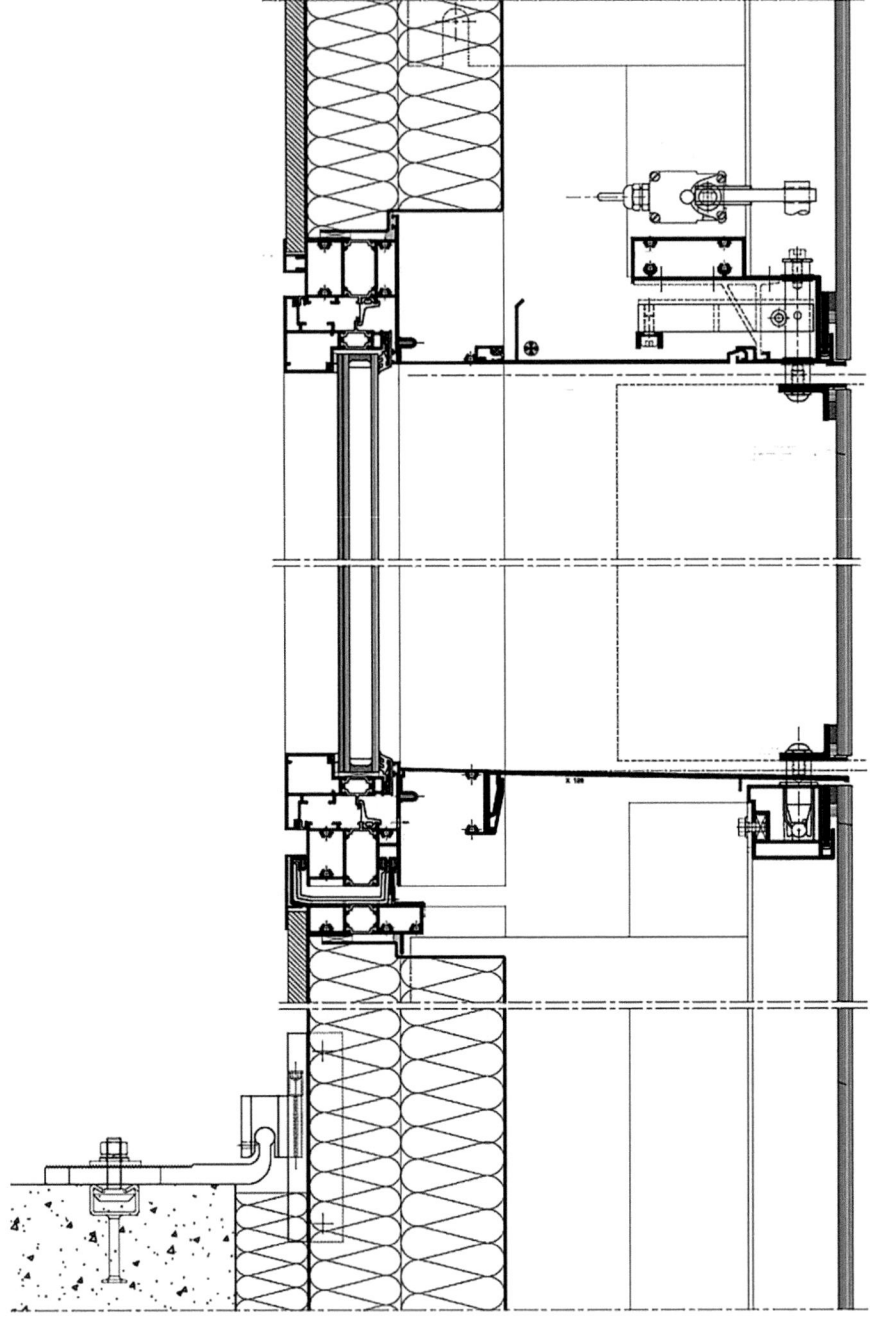

Fig. 3.13 Matthias Sauerbruch and Louisa Hutton, "Mac567" sector, *Maciachini Center*, Milan. Executive design of the double envelope system. The façade components are realized by the combined mullion profiles, in a section associated with the frames (fixed and opening), supporting the double-glazing panels and projecting towards the sunshade devices. © Courtesy of AluK

3.2 The Executive and Functional Constitution of the Combined … 71

Fig. 3.14 Matthias Sauerbruch and Louisa Hutton, "Mac567" sector, *Maciachini Center*, Milan. Application of the double envelope system. The assembly of the façade components involves engaging by means of fixing hooks (or "bayonets", connected to the mullion profiles) to the vertical studs protruding from the brackets (assembled to the slabs by means of the halfen profiles), taking over the *spandrel* section beyond the horizontal structures. © Courtesy of AluK

by its high energy-efficiency performance, achieved through the advanced façade system that distinguishes both the technological upgrading criteria and the aesthetic and functional aspects. The construction process involves the transformation of an existing building, equipped with a transparent cladding system, which is expected to have a significant impact on the comfort conditions of the interior office spaces. Within the architectural renovation, the new external surfaces exert a significant influence on the perceptive dynamics of the complex, accentuating the vertical dimension. Consequently, the architectural examination prioritizes the façade as the solution to ensure a high level of sun protection and the ability to view the interior and exterior environments from a dynamic and amplified perspective (Fig. 3.16).

The structural configuration of the architectural organism is determined by the interweaving of point elements and horizontal reinforced concrete elevations, upon which the prefabricated unit system components are assembled to constitute the

Fig. 3.15 Matthias Sauerbruch and Louisa Hutton, "Mac567" sector, *Maciachini Center*, Milan. Technical interfaces of the double envelope system. The mullion profiles of the façade components support the cantilevered bracket framing projecting towards the frames defined, in the intrados section, by the "U"-shaped box configuration for the passage of the adjustment joint of the rotatable panels and for the support of the aluminium sheet enclosure. © Courtesy of AluK

perimeter curtain wall. The façade (encompassing a total area of 1.900 sqm) is classified as a "thick surface", which, in accordance with the design of the external envelope, assumes the primary role of the architectural organism through:

- the order of the geometric texture, which, according to the modulation of the external glass "fins", is reflected in the morpho-typological development of the interior spaces, through which the vision of the surrounding environment is amplified, reflected and transformed;
- the extension of the variable modular pattern in the external glazed "fins". These fins act as brise soleil and are equipped with silk-screen printing that emphasizes the effect of light reverberation. This results in the blurring of the perimeter surface of the building, which is reminiscent of a "nebula", due to the multiplicity of reflections and transparencies.

3.2 The Executive and Functional Constitution of the Combined … 73

Fig. 3.16 Giovanni Vaccarini Architetti, *Société Privée de Gérance (SPG) Headquarters*, Geneva. Perceptive determination and amplification of the envelope, according to the geometric texture, modular pattern and light reverberation in the "augmented window" principle. The application of the double-skin façade typology allows for the containment of energy consumption, acoustic protection and passive ventilation combined with air conditioning. © Courtesy of Alex Filz

The envelope system is subdivided into prefabricated modules, which are aligned at each level along a horizontal band. This ensures spatial flexibility in the arrangement of internal partitions according to a regular pitch (interaxis span = 1.500 mm). The façade modules are determined by:

- the screen-printing process capable of producing a diaphragmatic effect, which functions as a shield against light radiation. This effect is achieved while maintaining the criteria of transparency and visibility to the outside environment;
- the capability to support thermo-hygrometric and lighting functions and to take on new expressive values through mediated forms of transparency;
- the compositive scheme of modulation by means of the glass blades, which offer the opportunity to identify multiple frames with singular viewpoints.

According to the concept design, the "thick surface", through the use of screen-printed glass and steel protruding supports, is proposed in a "volumetric" manner, constituting itself in the form of an architectural body, identified by means of:

- the perceptive "dematerialization" of the vertical enclosures and of the entire body, established in a diaphanous and ethereal form, sensitive to the luminous and chromatic variations of the surrounding environment;

- the vision in continuous mutation, progressively changing in appearance according to light incidence or darkness, as well as weather conditions.

The typological and functional articulation of the enclosure and diaphragmatic glass elements is aimed at improving the lighting quality of the interior spaces, also to the benefit of containing energy consumption. The low *thermal transmittance value* of the façade (equal to $U_w = 0.60$ W/sqm.K) is complemented by solutions designed to avoid any perimeter thermal bridge, where the sections of the horizontal elevation structures are covered by the insulating layer (made of rock wool; thickness $= 200$ mm) while the roof is protected by thermo-acoustic insulation (thickness $= 400$ mm). The envelope is defined in the unit system type with an aluminium frame (with dimensions of 1.500×3.150 mm) and with the division into both transparent and opaque panels: the components consist of modular elements with internal opening and external fixed wings, based on aluminium profiles with thermal break.

The unit system is installed floor by floor, based on the top-down construction method (i.e., from the roof to the ground floor), as opposed to the current practice of installing from the bottom-up. In particular, the aluminium profile frame, proceeding from the inside to the outside, supports:

- the insulating glass of the internal curtain wall, consisting of triple glazing with two cavities of the *Guardian Extraclear Tempered* float type (according to the values expressed by $FS = 50\%$; $RL = 16\%$; $TL = 67\%$; $U_w = 0.60$ W/sqm.K);
- the intermediate ventilated cavity, which includes the microperforated venetian blind device for light regulation (for the overall width $= 130$ mm);
- the extra-clear tempered HST single-glazing (thickness $= 10$ mm) of the external shading, defined by the *SunGuard Solar SiverLite 70 HD* type selective section (according to the values expressed by $FS = 69.5\%$; $RL = 26.9\%$; $TL = 69.6\%$; $R_w = 33$ db; $U_w = 0.60$ W/sqm.K).

The brise soleil are arranged with variable pitch (presenting basic dimensions of 200, 400 and 600 mm, for heights between 1.160 mm and 4.340 mm), while the screen-printing texture is based on a modular white pattern, the density of which grows from the inside to the outside following an irregular pattern (Fig. 3.17).

The executive design and the technical interfaces of the envelope components consider the overall strategy for the façade elements based on:

- the anchoring of the brackets aimed at supporting the external glass "fins";
- the construction of the naturally ventilated cavity to hold the upward convective airflows in the space between the two vertical surfaces;
- the interaction with the perimeter (underfloor) air-conditioning system.

The composition of the system regulates the axial modulation of the outer curtain and substructures, as in the case of the tubular steel profiles (enveloped by the double application of the fireproofing sheets, the insulating layer and the outer cladding). The axial arrangement of the vertical structures establishes the connection interfaces between the mullions of the double-skin components, according to:

3.2 The Executive and Functional Constitution of the Combined …

Fig. 3.17 Giovanni Vaccarini Architetti, *Société Privée de Gérance (SPG) Headquarters*, Geneva. Morphological and "volumetric" properties of the envelope through the contribution of luminous, perceptive and chromatic variations. Composite functional configuration of the different glass surface textures of the façade. © Courtesy of Alex Filz

- the combination of the extended triple tubular profiles (such that they cover the view of the vertical frames), provided with the double specular cavities for the insertion of the internal gaskets, on the axis of assembly, and the perimeter frame profiles, for the rebate of the opening sashes;
- the projection, perpendicular to the triple tubular profiles, of the frame with first extended tubular cavity and with outer tubular cavity supporting the frame profiles (with diagonal screw fastening) for the connection (by means of structural silicone) of the outer shading glass panels beyond the ventilated cavity;
- the insertion of the venetian blind device made of micro-perforated aluminium louvres in the cavity (Figs. 3.18 and 3.19).

The unit system components are assembled to the horizontal elevation structure through the application of the double welded plate bracket, through:

- the extrados flat steel plate, connected by dowelling;
- the vertical steel plate, connected to the outer vertical load-bearing section of the slab by dowelling, to which the hooked receiving device for engaging the "bayonet" profile (to the rear septum of the frame mullion) is jointed.

Moreover, the external section of the hook-receiving device is integrated with the double steel plate (connected by bolting), the joints of which intersect the *spandrel* section of the internal curtain wall: this forms the support surface for the overhang

Fig. 3.18 Giovanni Vaccarini Architetti, *Société Privée de Gérance (SPG) Headquarters*, Geneva. Methodology of construction of the external "fin" supports, the double-glazing enclosure and the cavity interposed between the internal curtain and the external screen. © Courtesy of Alex Filz

defined by the double horizontal plate and the vertical septum, to form the "C"-shape, which sustains the plate (connected by bolting in the slots that extend beyond the corners of the shape), from which the steel brackets continue to support the glass "fins". In particular, the brackets are arranged in the "forceps" type, where the bars are extended in different sizes for the assembly of "fins" of uneven base, according to the application by fixing point joints (within the holes produced in the thicknesses of the glass panels). The components of the façade are featured in the upper part by the projection of the *spandrel* section, supported by the double steel plate, which is directed both to the "bayonet" connection to the floor structure and to the support of the bracket for the external "fins": the stringcourse portion is enclosed by vertical double aluminium sheet cladding, with an insulating layer in between, and is connected to the horizontal double cavity frame profile with upper connecting wing (with rebate gasket on the outer cladding) and double cavity open for the insertion of direct connectors from the lower transom. This is composed of the tubular section, with the double side cavities mirroring the upper cavities, with the vertical partitions aligned and extended to form the frame profile for the triple-glazed opening frame rebate; subsequently, in order to facilitate the drainage of water, the extension of the crosspiece, with a double tubular cavity, is placed outwards in a lowered position to support the frame profile for the assembly (by means of structural silicone) of the external glass screen. At the same time, the lower

3.2 The Executive and Functional Constitution of the Combined … 77

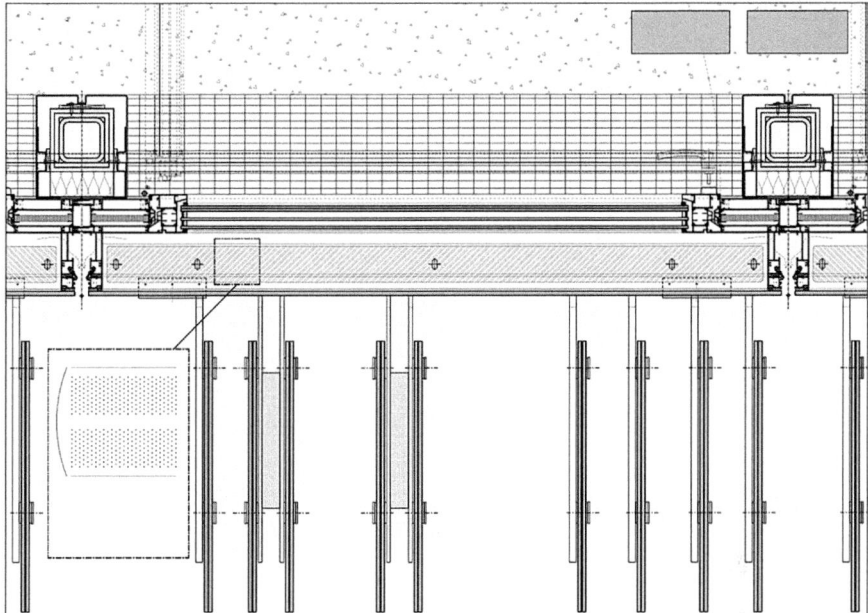

Fig. 3.19 Giovanni Vaccarini Architetti, *Société Privée de Gérance (SPG) Headquarters*, Geneva. Executive design of the mullion profiles with respect to the modular texture defined by the connections between the double-skin components, characterized by the extended tubular sections (covering the vertical frames) and supporting the external glass screen. © Courtesy of Pichler Projects

longitudinal septum of the outwardly projecting transom supports the sliding and adjustable venetian blind shading device within the gap (Fig. 3.20).

The components of the façade are specialized in the lower section through the insertion of the profile portion beyond the cladding elements of the *spandrel* band: this section is formed through the arrangement of the extended profile by means of the triple tubular cavity, up to the connection, beyond the floor extrados, with the profile of the opening wing. Externally, the lower part is aligned with the transom, which has a single tubular section with an external frame profile to support the external shading. The condition of the transom's alignment at a raised level is observed in order to allow the open band on the outside for convective airflows.

The assembly procedure for the brackets supporting the glass "fins" is employed for the intermediate horizontal levels, where the steel "C"-shaped support frames are connected to the blades stretched from the mullions: at the same time, in the upper and lower sections towards the supports, the aluminium crosspieces are spaced out in such a way as to continue the ventilation function of the cavity (Fig. 3.21).

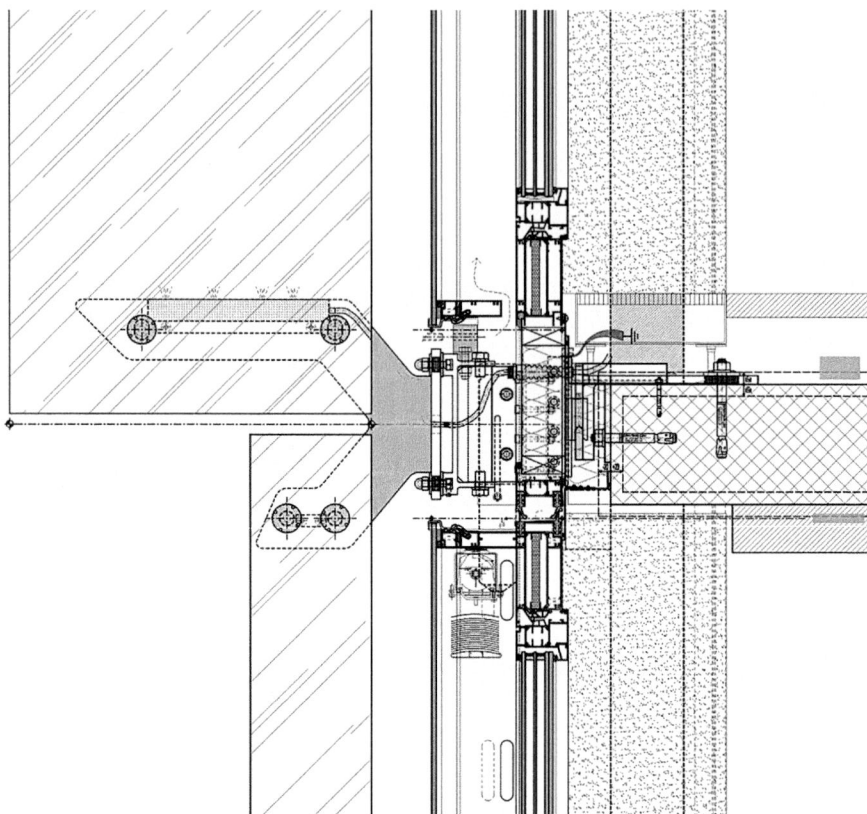

Fig. 3.20 Giovanni Vaccarini Architetti, *Société Privée de Gérance* (*SPG*) *Headquarters*, Geneva. Executive design of the double-skin unit system examined in relation to the horizontal elevation structure. The analysis takes place in an integrated manner with profiled connective procedures, and the inclusion of supports to the external glass "fins". © Courtesy of Pichler Projects

3.2 The Executive and Functional Constitution of the Combined … 79

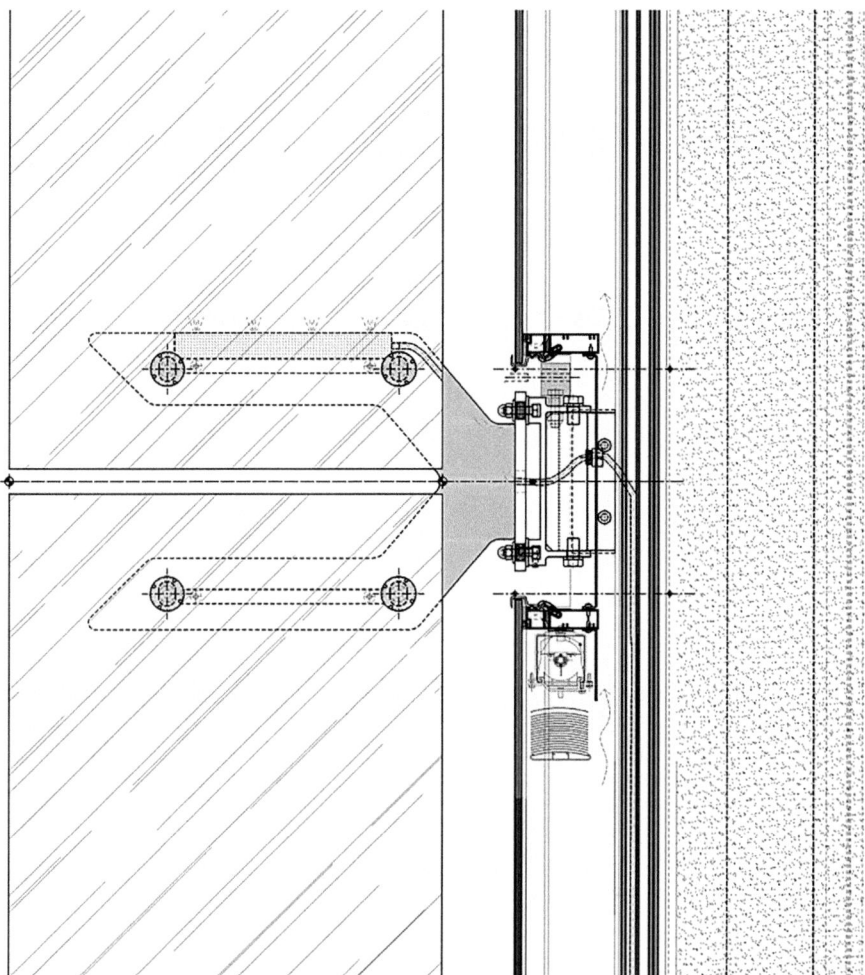

Fig. 3.21 Giovanni Vaccarini Architetti, *Société Privée de Gérance (SPG) Headquarters*, Geneva. Executive design of the intermediate assembly sections relating to the external glass "fins" to be conducted in accordance with the same connective procedure previously expressed in relation to the floor structures. The spacing of the stringers from the jointing device is to be maintained in order to ensure the ventilation of the cavity. © Courtesy of Pichler Projects

Chapter 4
The Technological Requalification by the Multi-layer Façade Systems

Abstract The study examines the executive and functional configuration of environmental and interactive façade systems, with a particular focus on ventilated multi-layer envelopes that integrate passive and dynamic mechanisms for climate and energy control. Central to these configurations is the double-skin façade, which activates natural convection through temperature differentials between the air cavity and external inputs, generating efficient vertical airflows. This process enables the modulation of thermal behavior across seasons: in summer, it facilitates convective cooling and limits solar gain; in winter, it retains heat and reduces dispersion. The envelope system is composed of prefabricated modular units integrating monolithic glass panels, thermally broken aluminium frames, automated shading elements, and partitioned cavities designed to reduce lateral air exchange while enhancing upward circulation. These façades function as environmental mediators, regulating light, temperature, and humidity to ensure interior comfort while optimizing energy performance. Their construction supports precise assembly, adaptability to varying building typologies, and facilitates microclimatic control via operable elements that respond to sunlight intensity and air movement. The internal cavity, further structured into autonomous vertical units, contributes to increase thermal inertia and acoustic insulation, while also facilitating moisture control and condensation reduction. The cell-like subdivision of the cavity supports the chimney effect and allows airflows to be calibrated module by module, improving aerodynamic behavior. Tailored profiles, gaskets, and mounting brackets enable the integration of shading slats and glazing panels, enhancing performance and ensuring mechanical continuity across the façade system. The operability of ventilation flaps, located near floor slabs, and the use of micro-perforated shading slats protected from weather, allow for passive air regulation based on solar radiation and temperature differences, ensuring responsiveness to diverse environmental conditions.

Keywords Double-skin façades · Passive climate control · Thermal convection flows · Modular envelope systems · Environmental sustainability · Energy-efficient design · Dynamic façade design

4.1 The Executive and Functional Constitution of the Environmental and Interactive Façade Systems

The technological re/qualification procedures through the use of multi-layer façade systems focus on the design of a ventilated advanced double envelope, in which the natural airflow triggers the convection process: this is due to the thermal gradient between the temperature in the cavity and the temperature of the air entering through the louvers, which creates a temperature difference. They are examined to ensure the correct size of the openings, and a careful examination of the dimensions, lower inlets, upper outlets and pressure losses is carried out to optimize the "chimney effect" phenomenon. The system's performance involves:

- the natural ventilation procedures for interior spaces, as well as controlling solar radiation in terms of both light and energy;
- the summer thermodynamic operative procedures whereby an increase in temperature within each module generates convective airflows that reduce the local temperature. In this case, the ventilated cavity acts against the phenomenon of heat loss through natural convection by generating an upward airflow through the heating of the outer surface of the glass screen: therefore, the airflow results in a lower temperature inside the cavity than outside, which reduces the amount of heat that can be absorbed by the internal enclosure;
- the thermodynamic winter operative procedures involve storing heat inside each module to mitigate the thermal gradient between the inside and outside, thereby reducing heat loss. In this case, heat loss is reduced during periods when the outdoor temperature is lower: this reduces the amount of heat that comes in through solar radiation and speeds up the movement of water vapour out of the building. In addition, the double envelope system's constitution also prevents:
- the thermal accumulation (i.e. heat transmitted from internal spaces to the outside) by allowing ventilation in the cavity, which keeps the average temperature of the internal enclosure near that of the internal spaces;
- the reduction of condensation, facilitating the evacuation of water vapour from interior spaces and enabling the disposal of humidity due to water infiltration.

These principles are embodied and expressed in the façade system designed by Hof Associati for the redevelopment of *Palazzo Grossi* as the new headquarters of the Municipality of Perugia's Technical Services Centre. The design objectives of preserving the building's identity, which is deeply associated with the collective memory, and the original courtyard plan, focus on enhancing the quality and environmental well-being of the interior spaces. This is achieved through the implementation of a combined bioclimatic control system comprising a roof garden, a water pool and coloured glass, and a double-skin façade (conceived in collaboration with Schüco). The design of the double-skin façade is based on a consolidated expression of scientific and practical knowledge, technology transfer procedures, and methods for designing components, all of which stem from the evolutionary study of technical skins. The project proposes to comply with the requirements of the new use while preserving the original typological composition and the collective memory of the building. It is organized into a series of strategic actions, the most prominent of which are the restoration or reinstatement of the external façade and the visual containment of the extension works inside the courtyard (Figs. 4.1, 4.2, 4.3 and 4.4).

The latter are conceived as an intrusive new addition, implanted in the soft-tech courtyard. The redevelopment confirms the original courtyard typology, as it represents a precious historical heritage and constitutes the ideal setup to provide the necessary environmental quality. This is enabled by the ventilation and direct lighting of the main distribution corridors and work spaces, which allows for microclimatic control conditions to be created by adjusting the windows, doors and sunlight intensity. The façade consists of modules, each composed of an external monolithic glass panel, a motorized shading device made of micro-perforated slats, and a mullion and transom curtain wall behind, with opening and fixed parts, made of thermal break aluminium profiles and double-glazing (with high insulating capacity and low light energy reflection). On each floor, the space between the two enclosures is divided into five units, each of which is insulated by monolithic glass partitions to minimize airflow in the transverse direction. Each module is autonomous in the vertical direction and is equipped with finned panels for air inlet and outlet, as well as flow conveyors located behind them and at the floor slabs. The visual confinement is obtained by the upper crown of corten steel, which reduces the visual impact and acoustic stress emitted by the technical plane. Beyond the planar surface of the glass-shielded façade, the architectural apex of the building emerges in a gesture that reconnects with the context's massive character (Figs. 4.5 and 4.6).

The bioclimatic function of the double envelope system on the courtyard façade, aimed at optimizing the climatic comfort inside the spaces, is combined with the creation of a roof garden. It contributes to the micro-environmental control of the site through the extensive green roofing, which is intended to ensure the improvement of the climate by decreasing the temperature, which impacts the surrounding buildings. The double envelope system limits energy consumption, through the reduction of the temperature range and the thermal insulation effect (Figs. 4.7 and 4.8).

The construction of the system takes place beyond the perimeter structural apparatus (in circular reinforced concrete pillars), according to the spatial and dimensional order that coordinates the modular and equipped partition walls on the inside. The double-skin façade is a prefabricated unit-type production and construction system for on-site assembly. It is based on the attachment brackets on halfen profiles, which are embedded in the concrete casting and connected to the steel reinforcements. The experimental system is realized according to:

- the development of the frame to support the internal double-glazing enclosure and the extension to support the external screen. Specifically, the thermal break frame profiles for the internal curtain wall comprise the main rear section for the mullions and transoms. This is established by the rectangular intermediate chamber and the open chamber, which acts as the insertion seat for the internal vertical and horizontal gaskets. This section features a cavity and ribs for the

Fig. 4.1 Hof Associati, *Palazzo Grossi*, Perugia. Morpho-typological identity and innovative building systems: the redevelopment of the existing building involves combining enhanced and upgraded building components, functional processes, and envelope systems while maintaining the original typological configuration. © Courtesy of Hof Associati

4.1 The Executive and Functional Constitution of the Environmental …

Fig. 4.2 Hof Associati, *Palazzo Grossi*, Perugia. Poetic and executive strategy: the intervention combines the use of bioclimatic models, which are expressed according to the sinergic contribution of environmental grafts and equipment in the double envelope façade system, with the aim of calibrating thermo-hygrometric and lighting conditions inside. © Courtesy of Hof Associati

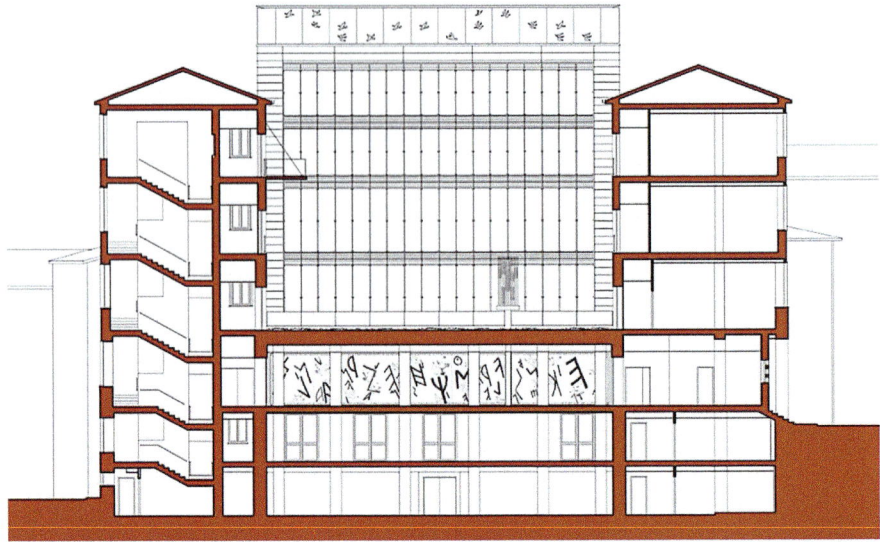

Fig. 4.3 Hof Associati, *Palazzo Grossi*, Perugia. Morpho-typological development: the spatial and connective interfaces bring together two parts of the building, with the mezzanine level acting as a unifying element. This includes the incorporation of a double envelope façade system and closing and covering curtains, which are characterized by artistic applications and graphic decorations on the entrance glazing and the roof apex wrapping. © Courtesy of Hof Associati

Fig. 4.4 Hof Associati, *Palazzo Grossi*, Perugia. Typological configuration: the development of the courtyard contributes to the environmental and bioclimatic control of interior spaces by providing ventilation through the convective flows towards the façade. © Courtesy of Hof Associati

4.1 The Executive and Functional Constitution of the Environmental … 87

Fig. 4.5 Hof Associati, *Palazzo Grossi*, Perugia. Bioclimatic functioning of the envelope system: the façade system, made of modular components (with monolithic glass outer panels, shading devices in the cavity and internal enclosures with opening and fixed parts), involves the subdivision into insulated cells by means of monolithic glass partitions. © Courtesy of Hof Associati

installation of the "L"-shaped profile, which allows for the inclusion of internal gaskets on the double-glazed panel;
- the composite outer front section, connected to the previous one by a couple of polyamide bars and characterized by:
  - the profile section with an open chamber that acts as an insertion seat for the outer vertical and horizontal gaskets;
  - the wing extension towards the external gaskets on the double-glazed panels;
  - the projection of the "L"-shaped profile section provided with a double partition for the insertion of support elements for external shading and a perpendicular wing with a cavity for a central gasket. This wing also serves as the point of fixing for the shaped profile designed to hold the glass blade perpendicular to the façade plane, and separates the ventilated cavities for each module.

The screens are connected to the clamping elements, which are inserted and locked within the tubular section that protrudes from the outer frame profile. These elements extend towards the outer end, where they support the glass panel by means of transverse bolting. In particular, one of the brackets on each side allows the assembly of the "L"-shaped profile with the insertion seat for the panel. Overall, the design and

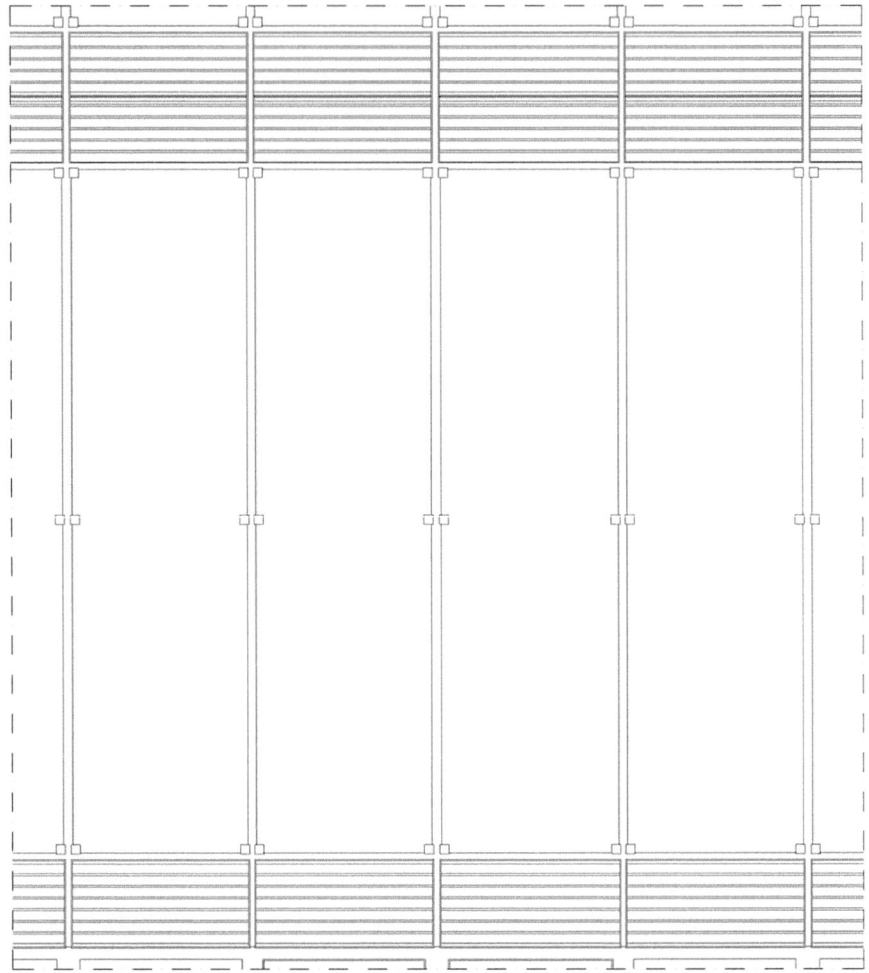

Fig. 4.6 Hof Associati, *Palazzo Grossi*, Perugia. Modulation of the system: the geometric and spatial composition of the façade units is planned according to the principles of natural ventilation and solar input control, by reducing the thermal gradient to prevent heat accumulation (in winter) and ensuring the envelope is naturally ventilated (in summer). © Courtesy of Hof Associati

production of the unit components is determined by the execution of the aluminium load-bearing frame and the fastening elements, specifically comprising:

- the brackets for fixing to horizontal structures, manufactured from stainless steel and in the form of "L"-shaped corner elements. They are connected directly to the vertical side partitions of the mullions. These profiles enable connection to the internal plates, which are attached to the brackets on the slabs;
- the upper brackets for fixing the grid, inserted into the cavity, and the fixing brackets for the venetian blind support devices;

4.1 The Executive and Functional Constitution of the Environmental … 89

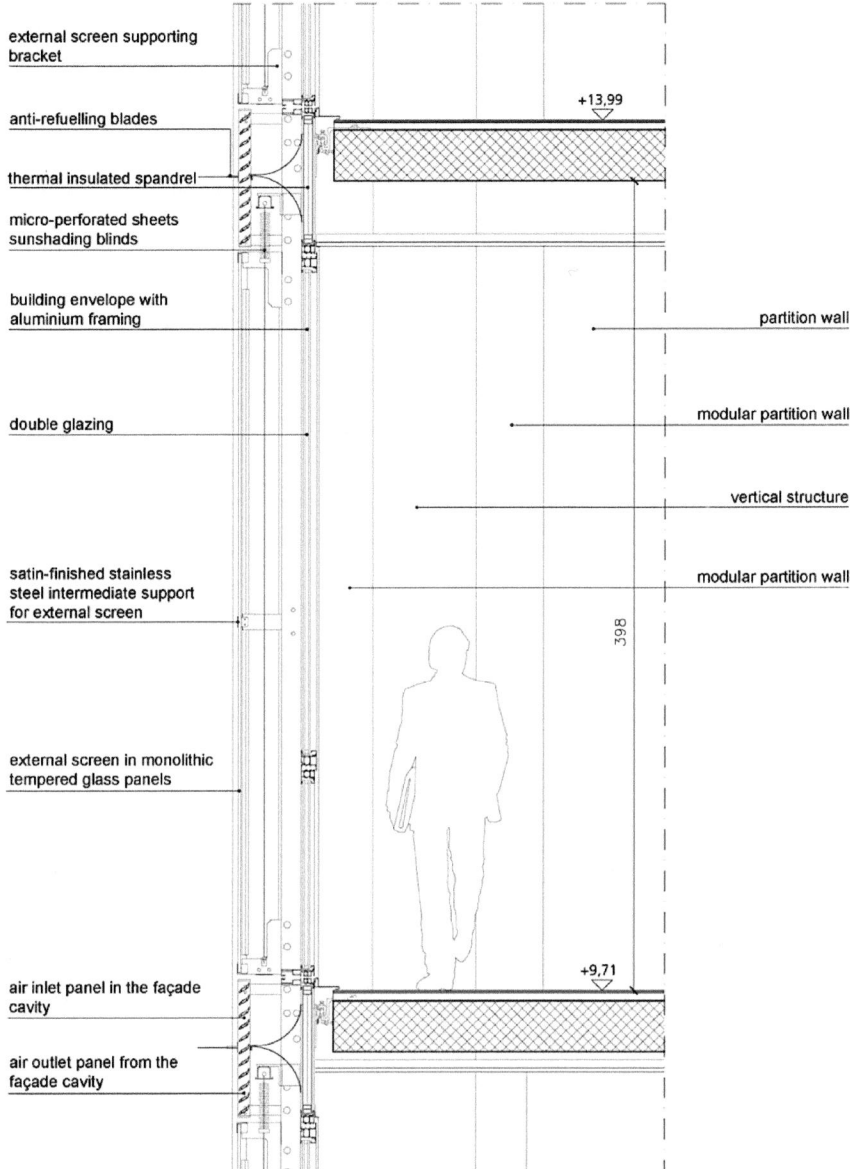

Fig. 4.7 Hof Associati, *Palazzo Grossi*, Perugia. Ventilation process of the envelope system: the double-skin façade detects the application of the aerodynamic device for conveying and releasing convective airflows in the cavity, using deflector profiles divided by curved shapes designed to accompany the upward flows of passive ventilation. The enclosure system, supported by aluminium mullions and transoms, involves, on the inside, the assembly of profiles to support the double-glazed panels with an opening section, and, on the outside, the monolithic glass screen suspended from the structure by means of steel brackets and fasteners. © Courtesy of Hof Associati

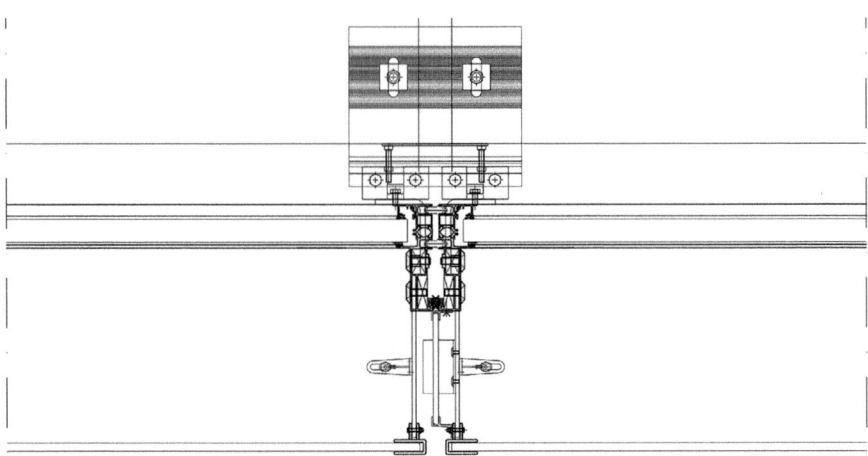

Fig. 4.8 Hof Associati, *Palazzo Grossi*, Perugia. Conception of the ventilated double-skin façade based on advanced screen: the technology involves triggering the natural ventilation process by the finned louvers to create an upward airflow. © Courtesy of Hof Associati

- the side fixing brackets for the glass screen, with a cantilevered geometric configuration and a satin-finish stainless steel retaining device, applied in three positions (upper, middle and lower);
- the lower fixing brackets for the glass screen, with a linear geometric configuration and a satin-finish stainless steel retaining device (Figs. 4.9, 4.10 and 4.11).

The composition of the system provides for the partitioning of the cavity between the unit components by means of monolithic glass panels, with the aim of reducing transversal ventilation and, therefore, enhancing vertical convective airflows (according to the "chimney effect" operating). The system is thus delineated in the "cell" configuration by units that are independent of each other on a functional level. The passive functioning requires the ventilated cavity to perform integrated functions to achieve complex mechanisms of dynamic interaction with external climatic conditions: these are either permanent (e.g. to increase thermal inertia and acoustic insulation related to the internal curtain) or temporary (e.g. to dispose of water vapour in internal spaces during periods of reduced environmental temperature, or to cool these spaces during periods of high temperature) (Figs. 4.12 and 4.13).

The ventilation flaps are positioned on the exterior of the modules, next to the horizontal structures. This allows air to circulate between the cavity and the interior spaces, and in the opposite direction. The project specifies that each module is autonomous, with air intake and exhaust finned panels located at floor slab level (in accordance with the box-windows façade type). The two aerodynamic sections are divided by curved shapes engineered to guide convective airflows within the cavity, which also incorporates micro-perforated sheet metal blinds. In this way, the system acts as a ventilated cavity, providing passive aeration and facilitating airflow from outside to inside. The sunscreen in the cavity, for which the positions of the supports

4.1 The Executive and Functional Constitution of the Environmental … 91

Fig. 4.9 Hof Associati, *Palazzo Grossi*, Perugia. Profile configuration: frame to support the internal glazing and extension to the frame to support the external shading. Façade component modulation: internal curtain with fixed and opening elements. © by the Authors

are established, is manufactured from adjustable shading slats of micro-perforated sheet metal, protected from the weather and external pollutants: the shading reduces heat input depending on the external temperature and solar radiation conditions, being particularly effective when the outside temperature is lower than the interior space temperature and when overall radiation values are low (Fig. 4.14).

The production processing of the system modules is realized through the coordination of the profile sections and their connective interfaces. This is based on the fine-tuning and analysis of the part supply and manufacturing codes. This process takes over the handling of both the overall geometric dimensions and the analysis of the cutting, part assembly and bracket engage dimensions (Figs. 4.15, 4.16, 4.17, 4.18, 4.19, 4.20 and 4.21).

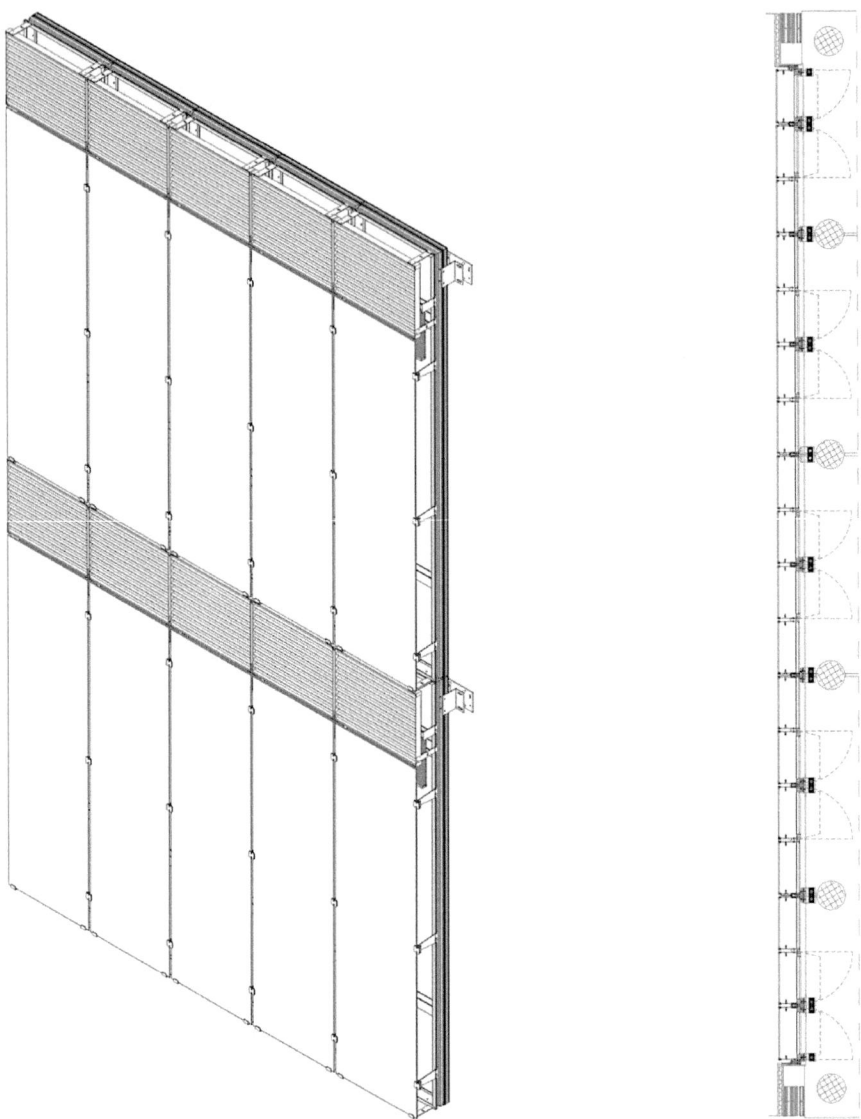

Fig. 4.10 Hof Associati, *Palazzo Grossi*, Perugia. Production and construction process: study of the double-skin façade system using frame profiles from the standard production model, which can be attributed to the single curtain walls series. The process involves the use of special moulds to create reinforced elements that support the external glass screens. © Courtesy of Hof Associati

Fig. 4.11 Hof Associati, *Palazzo Grossi*, Perugia. Environmental devices, technical interfaces and supporting/fixing brackets of the double-skin façade. © Courtesy of Hof Associati

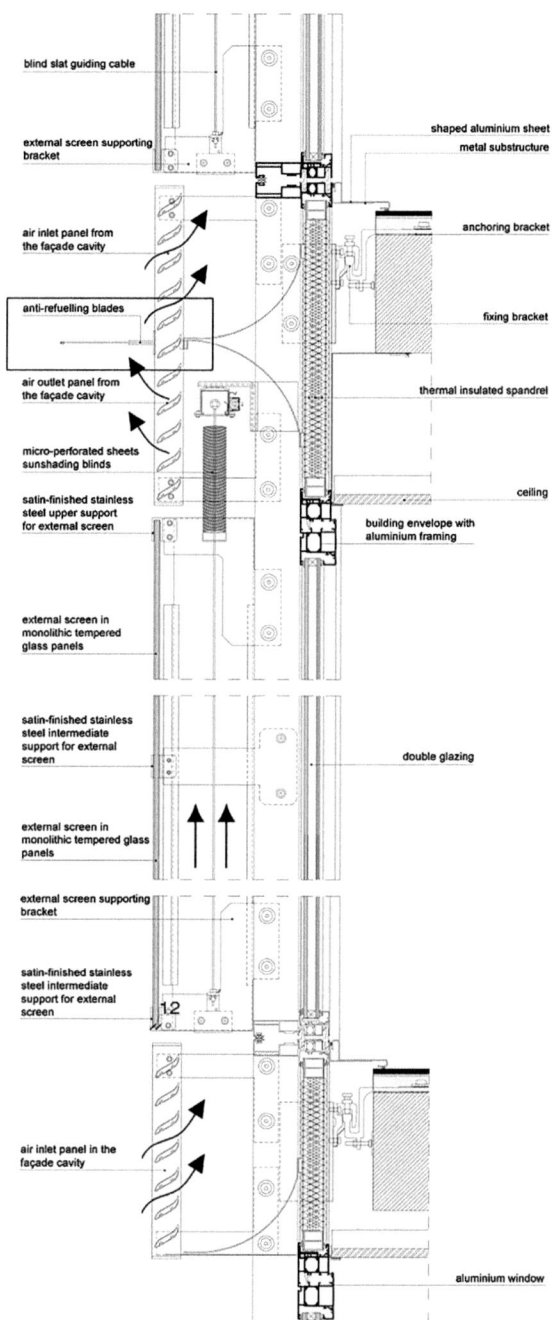

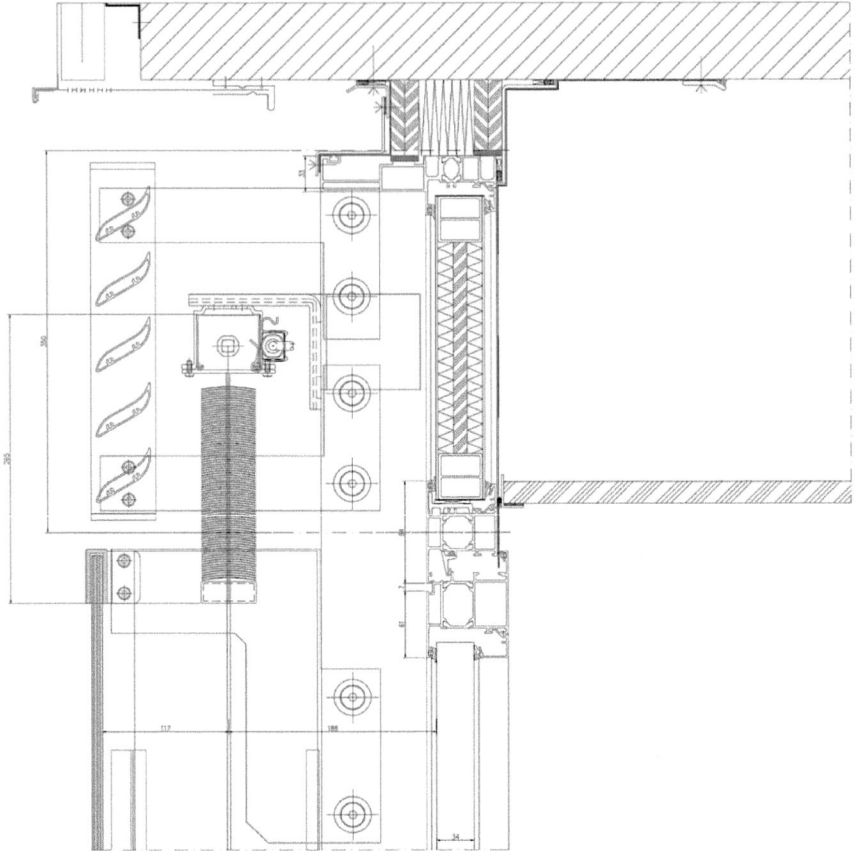

Fig. 4.12 Hof Associati, *Palazzo Grossi*, Perugia. System composition: the external screen is manufactured from tempered monolithic glass panels, allowing the ventilated cavity, or *buffer zone*, which provides thermal storage and insulation. This reduces wind pressure on vertical surfaces of window and door frames related to the curtain wall. © Courtesy of Hof Associati

4.2 The Architectural and Performance Reconfiguration by Technological Substitution of Envelope Systems

The procedures for re/qualification, as applied to façade curtains, are intended to develop building types with renewed expressive, functional and interactive features. The transformation of the perimeter plan involves designing buildings that serve different purposes and are easily identifiable. This is achieved by applying a multiple-skin façade in accordance with the different levels of solar exposure, in order to optimize energy performance. In addition, re/qualification procedures consider the dynamic interaction with thermo-hygrometric, light and air stresses. These flows can be adjusted and incorporated into the overall functioning of the system: in this way,

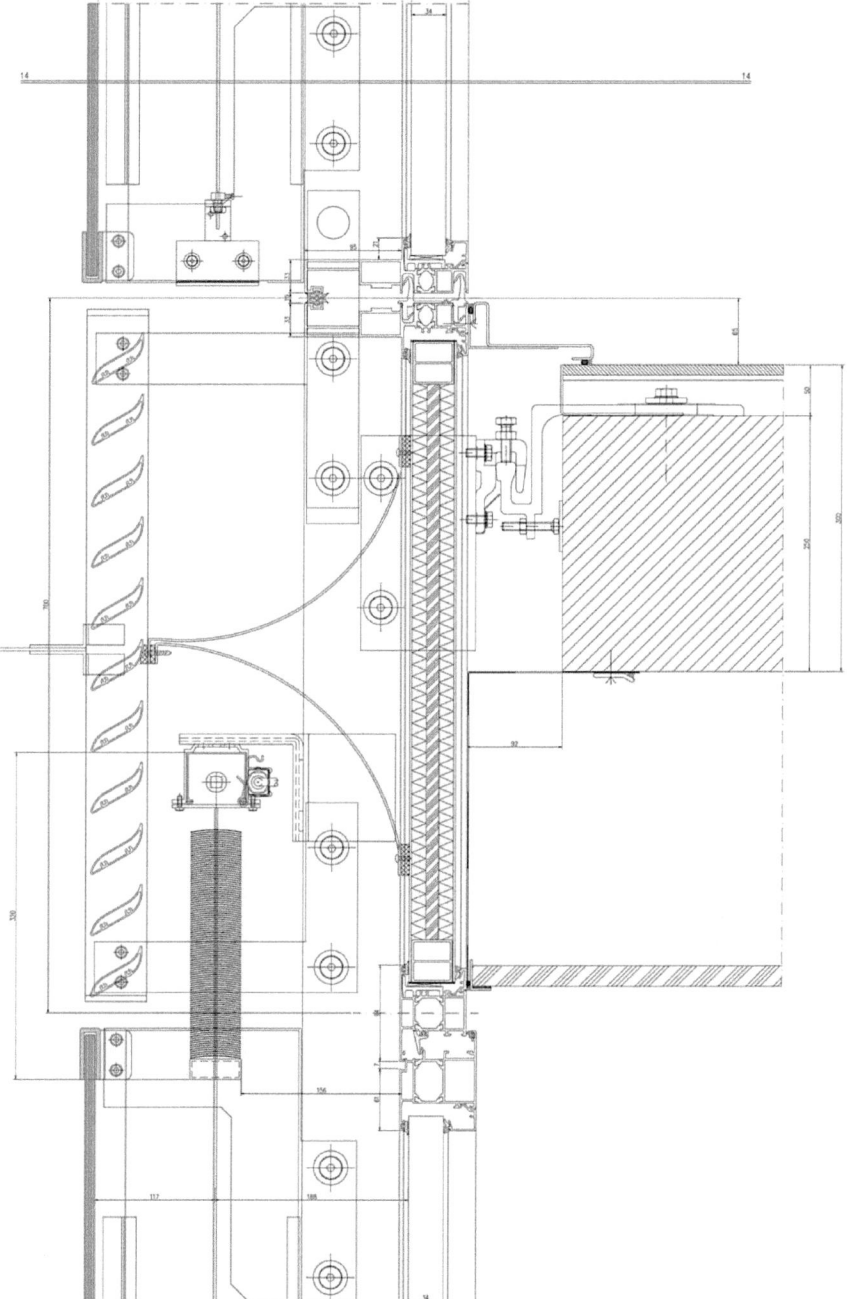

Fig. 4.13 Hof Associati, *Palazzo Grossi*, Perugia. Mechanical assembly procedures for double-skin prefabricated units with regard to extrados brackets. © Courtesy of Hof Associati

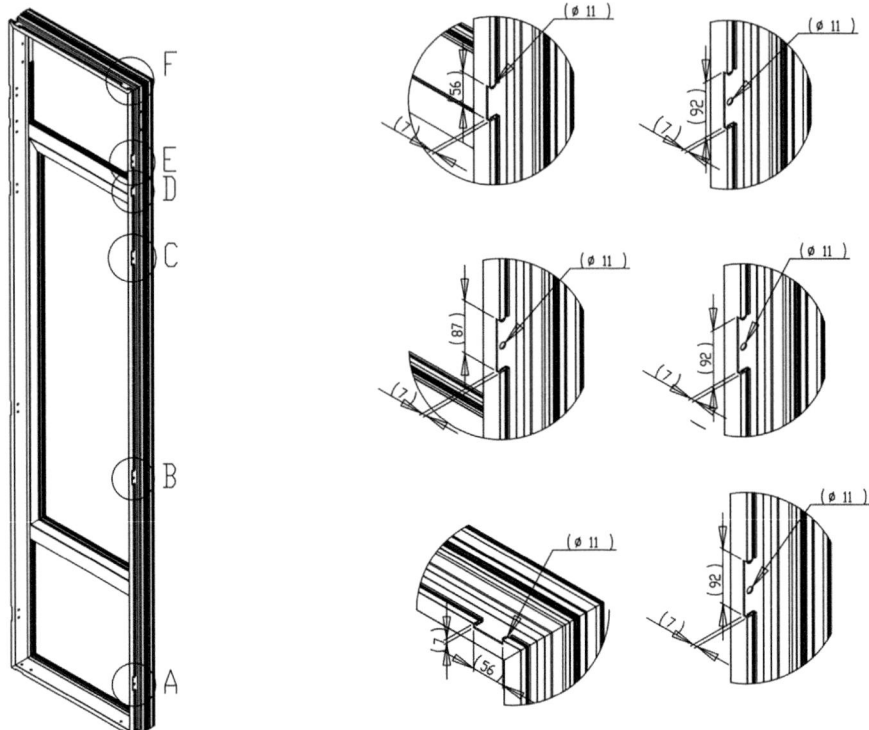

Fig. 4.14 Hof Associati, *Palazzo Grossi*, Perugia. Production drawings: processing of the system modules with respect to the cuts on the frame sections directed to the connection of the fastening brackets to the horizontal structures, of the upper brackets for fastening the grid to the cavity, of the fastening brackets for the supporting devices of the venetian blind screen, of the lateral and lower fastening brackets of the glass screen. © Courtesy of Hof Associati

upgrading procedures consider the tuning of unit systems as mediating (i.e. reflecting, capturing and diffusing) equipment against climatic stresses. The procedures for the redevelopment of building envelope systems are developed with receptiveness to adaptation and reorganization of:

- the typological arrangements through functional layering and sector-based overlapping to accommodate multiple activities;
- the possibilities for *eco-efficient* use of non-renewable energy resources, together with the acquisition, transformation and re-use of polluting emissions.

This research examines the redevelopment of the *Garibaldi Towers* in Milan as a contemporary case study. The project involved the conservative refurbishment of two office towers near Porta Garibaldi Station (each 23 storeys high) to accommodate new functions. The morpho-typological elaboration directed by Progetto CMR was arranged within a framework of defined and pre-existing conditions. The architectural, productive and executive implementation of the façade components,

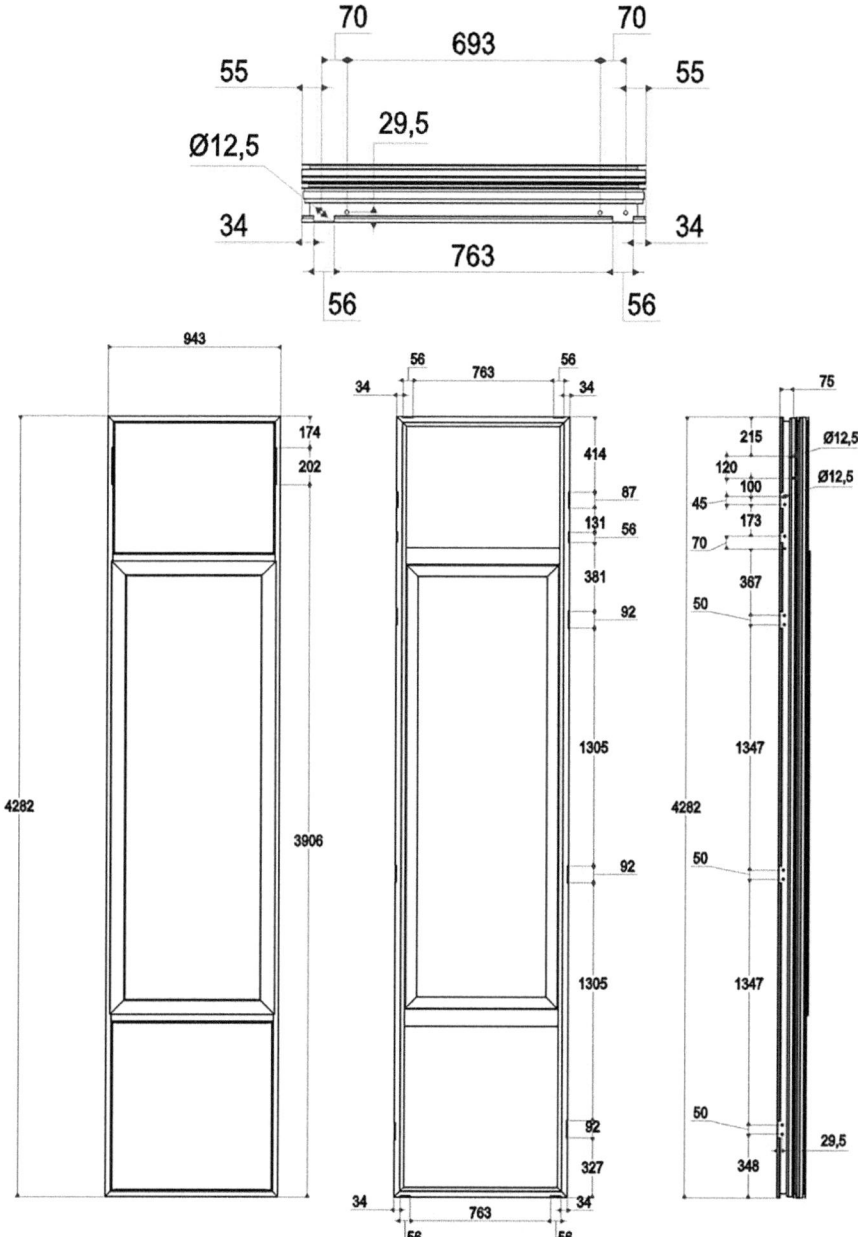

Fig. 4.15 Hof Associati, *Palazzo Grossi*, Perugia. Production drawings: processing of modules according to the coordination of sections and connective interfaces. © Courtesy of Hof Associati

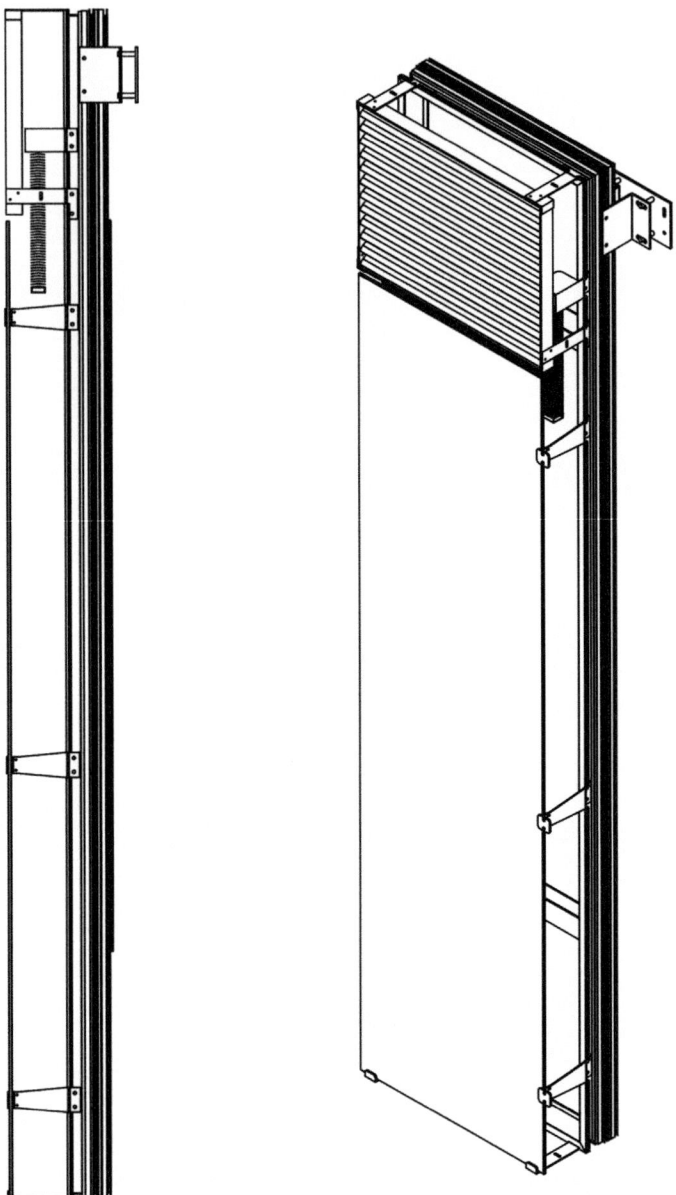

Fig. 4.16 Hof Associati, *Palazzo Grossi*, Perugia. Production drawings: frame supporting the internal glass enclosure and extension of the frame supporting the external shading. Modulation of façade components: internal curtain with fixed and opening elements. © Courtesy of Hof Associati

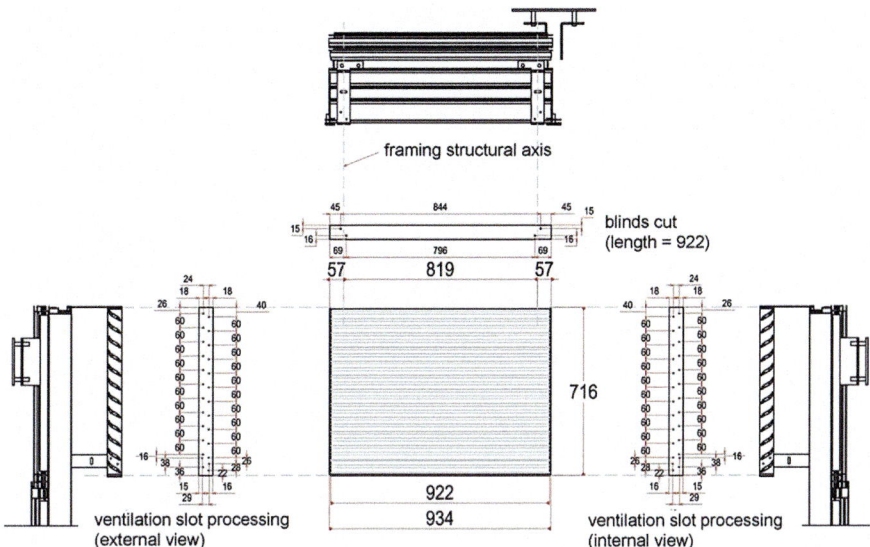

Fig. 4.17 Hof Associati, *Palazzo Grossi*, Perugia. Production drawings: ventilation slots geometric and dimensional modules according to the assembly to the brackets. © Courtesy of Hof Associati

Fig. 4.18 Hof Associati, *Palazzo Grossi*, Perugia. Building sequences: application of the fixing brackets to the horizontal structures and connection to the internal plates. © by the Authors

Fig. 4.19 Hof Associati, *Palazzo Grossi*, Perugia. Building sequences: application of the cantilever brackets protruding from the internal mullions to support the ventilation slots. © by the Authors

Fig. 4.20 Hof Associati, *Palazzo Grossi*, Perugia. Building sequences: application of the cantilever brackets protruding from the internal mullions to support the external screens. © by the Author

4.2 The Architectural and Performance Reconfiguration by Technological …

Fig. 4.21 Hof Associati, *Palazzo Grossi*, Perugia. Building sequences: procedures for the external screen according to the assembly carried out with traditional scaffolding. © by the Author

which characterizes the east and west elevations, is realized through the adoption of the prefabricated and independent unit system type, through the curtain walling in modules with double-glazed panels (with openable internal shutters), the ventilated cavity (which can be inspected) and the external screening in laminated glass.

The unit system components, which are based on vertical rectangular geometries, are formed by the load-bearing structure. This structure is made of thermally broken aluminium profiles which support the external frame of variable thickness. This frame is capable of providing the different facets on the curtain plane. The cavity, which is inserted between the curtain wall section and the shielding and contains the parietodynamic type of operation with forced ventilation, is proposed with both variable geometries and dimensions on the four vertices (between the outer panels and the inner insulating panels). The *spandrel* type insulating panelling, consisting of the external steel sheets, the insulation with mineral wool and the internal side in galvanized steel sheet (which turns horizontally up to the external glass), is placed inside the cavity, in the inter-floor area. Then, the façade includes internal framing to support the opening sash for cavity inspection and maintenance. These consist of double-glazed enclosures with laminated inner safety glass panels. The opaque parts of the ventilated façade consist of modular composite elements with a sand-blasted, light, natural stone cladding applied to a steel substructure. The aluminium honeycomb support lightens the panel. The envelope system with respect to southern exposure is realized through the enclosures into which the bioclimatic "greenhouses" and external sunshade devices are inserted.

The system's executive constitution foresees the use of a continuous steel UPN beam profile at the perimeter interface of the horizontal elevation structure:

- with respect to the upper wing, this involves the connection surface of the angular steel element that contains the integrative concrete casting above the self-supporting corrugated sheet metal (which is in turn realized above the steel frame). In particular, the concrete casting section is further restrained by the steel plate that is welded to the vertical wing of the corner element;
- with respect to the ends of the wings, the steel plate is welded on to act as a support plane towards the steel angle bracket.

The construction of the unit system components takes into account the engagement of the joint pins above the horizontal wing of the brackets through the assembly, which is achieved by combining the "bayonets" with the frame mullions. The perimeter section shows how the assembly strip is closed off between the horizontal structure and the components using two aluminium sheet extensions. These extensions contain both the fireboard panels (as vertical and horizontal glass fiber-reinforced plaster fire-resistant) and the thermal insulation layer (Figs. 4.22, 4.23 and 4.24).

Fig. 4.22 Progetto CMR, *Garibaldi Towers*, Milan. Visualization of the first building subjected to envelope system replacement and the second building with perimeter cladding in traditional prefabricated panels before upgrade procedures. © Courtesy of Progetto CMR

Fig. 4.23 Progetto CMR, *Garibaldi Towers*, Milan. Visualization of both buildings subjected to envelope system replacement by the same multi-layer façade. © Courtesy of Progetto CMR

The façade components oriented to the upgrading, as technological resilience tools capable of activating performances of absorption and reaction towards the perturbative pressures caused by the multidimensional obsolescence phenomena, with respect to existing building structures and sections, are defined in relation to:

- the lower and upper interfaces, by the profile geometries of the transoms being laid in accordance with the interposition of the wing elements, ribs and gaskets;
- the profile sections supporting the double-glazing enclosures;
- the pronounced portion on the outside, of box-shaped composition and of variable dimensions in function of the curtain at inter-floor level, to support the laminated glass screen, in accordance with the geometry of the cavity thickness.

Specifically, the upper part of the components includes the addition of a *spandrel* section, equipped with a thermal insulation layer (vertical) and a double fireboard panel enclosure. Furthermore, the composition of the cavity located between the perimeter enclosures and the external screen includes supports for the sunshade blades, with the upper portion covered by the internal cladding (Fig. 4.25).

The rearward positioning of the façade involves the insertion of the horizontal structure consisting of HE steel profiles, included in the dimensions of the main beam and supporting, via the lower wing, the connecting brackets to the aluminium frame for the installation of double-glazed enclosures: in particular, the assembly thickness of the enclosures, for the upper fixing, connects to the extension of the

Fig. 4.24 Progetto CMR, *Garibaldi Towers*, Milan. Faceted composition and heterogeneous external surface of multi-layer façade systems equipped with interactive cavities designed to protect and calibrate external loads in an *eco-efficient* manner (with respect to morpho-typological, energy and environmental re/qualification objectives). © Courtesy of Progetto CMR

Fig. 4.25 Progetto CMR, *Garibaldi Towers*, Milan. Building sequences: technical interfaces and assembly procedures of the independent façade components in the prefabricated unit type, on the perimeter steel frame jointed to the existing horizontal structure. © Courtesy of Progetto CMR

lower enclosure panels. In the case of a recessed façade plan, the string course section is marked by the lower projection (of the *spandrel* type) characterized by:

- the configuration of the extrados transom to the geometry aimed at completing the wing extensions and ribs, turning towards the plan surface of the frame capable of supporting multiple strips of fireboard type panels;
- the configuration of the intrados transom according to the geometries and dimensions aimed at supporting both the insulating layer and the vertical fireboard panel, allowing for connection, jointing and levelling with the sheets supporting the lower enclosure panels;
- the external perimeter steel cladding, which is in turn concealed by the extension of the laminated glass screen (Figs. 4.26 and 4.27).

The composition of the side interfaces is set by the wings of the perimeter steel pillars, with respect to the planar arrangement of the curtain wall, creating the support

Fig. 4.26 Progetto CMR, *Garibaldi Towers*, Milan. Building sequences: technical interfaces and assembly procedures of the independent façade components according to the staggered and irregular development of the external screens. © Courtesy of Progetto CMR

plane for the brackets protruding towards the connection devices connected to the mullions. These are defined by:

- the outer side section that accommodates the double vertical sheet metal connecting strip, which holds the fireboard panels and the thermal insulation layer in place. Moreover, the external surface of the fitting allows the thermal insulation layer (extending from the adjoining component) to be butted up against it and the shaped cladding section to be extended up to the structural frame towards the laminated glass screen;
- the inner side portion directed towards the support of the frame profiles (which can be opened) supporting the double-glazed enclosures (Figs. 4.28, 4.29 and 4.30).

The intermediate interfaces are based on the combined and symmetrical articulation of the mullions, conceived with respect to the internal portion, which provides support for the opening frame profiles, the extension of the side partitions, including

4.2 The Architectural and Performance Reconfiguration by Technological … 107

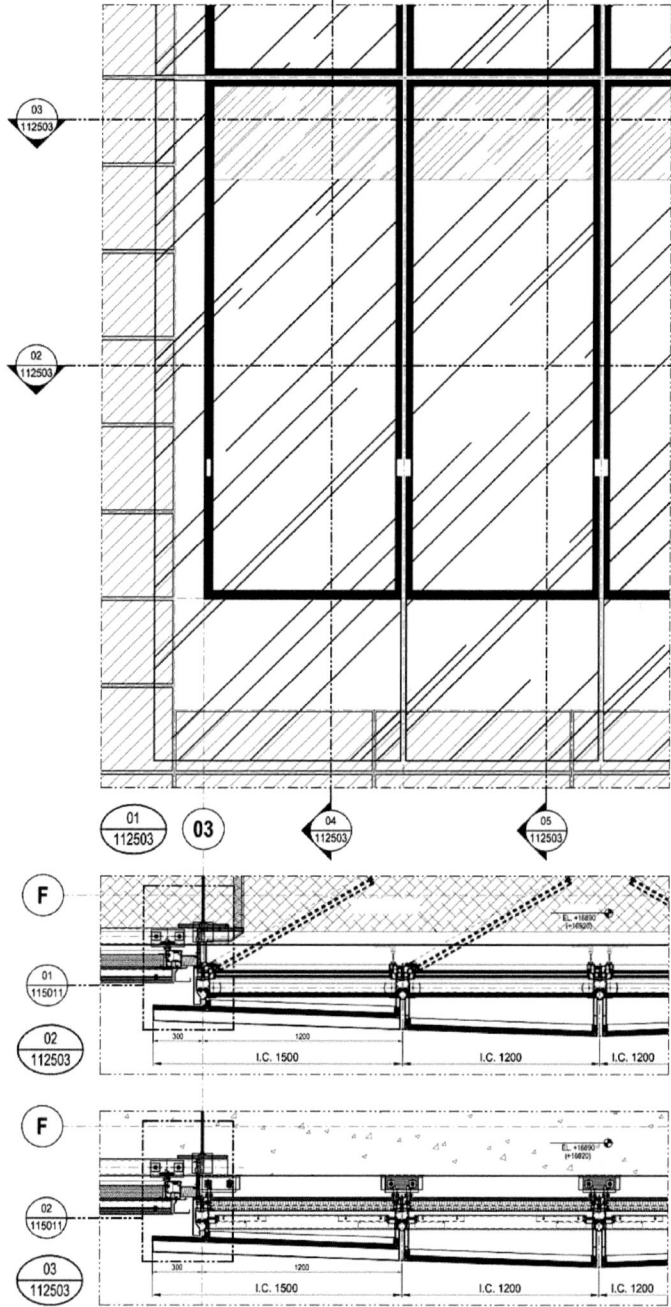

Fig. 4.27 Progetto CMR, *Garibaldi Towers*, Milan. Prospective and constructive constitution of multi-layer **façade** components (with internal openings) according to mechanical assembly methods to perimeter brackets and side connections between mullions. © Courtesy of Progetto CMR

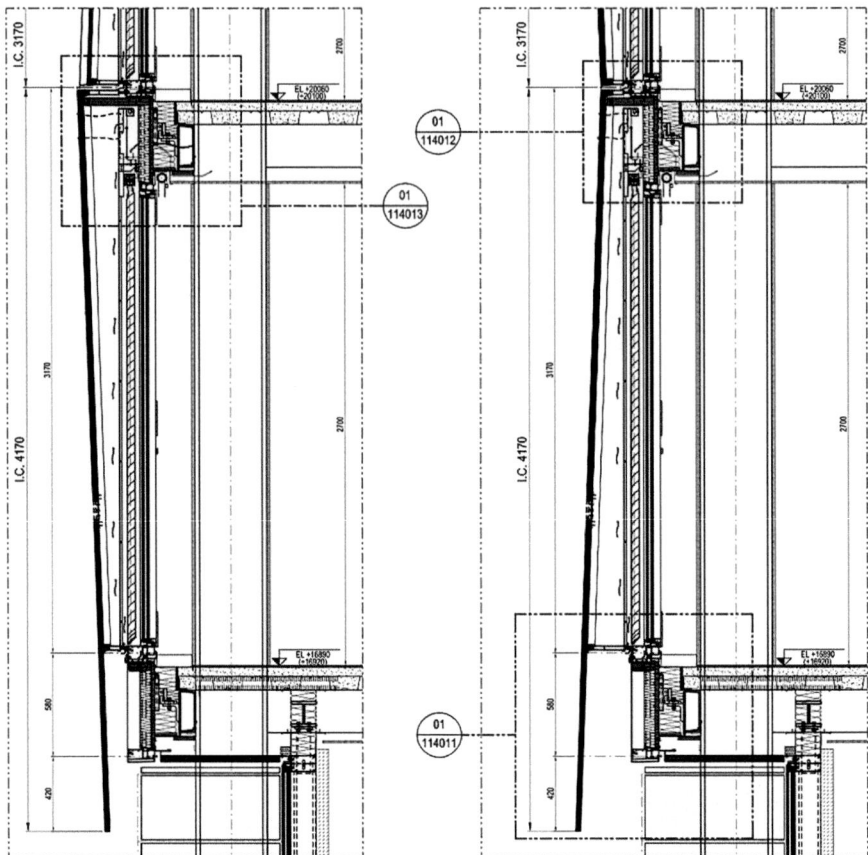

Fig. 4.28 Progetto CMR, *Garibaldi Towers*, Milan. Development of technical interfaces related to the construction of frames and connection devices to the main elevation structures; detection of joints to the extent of the cladding and insulation projections. © Courtesy of Progetto CMR

the stiffeners required for the assembly of the sunshade, and the box-shaped projections directed towards the segmented succession of external screens. The type of adjoining multi-layer component is determined on the basis of:

- the connection of the fitting oriented towards the shaped side sheet metal belonging to the prefabricated façade unit;
- the execution of the brackets related to the perimeter frame, through assembly using "bayonets" combined with mullions with a box-shaped section: this allows the support elements to be attached to the multi-layer section (including the insulation panel) and, on the outside, to the cladding surfaces.

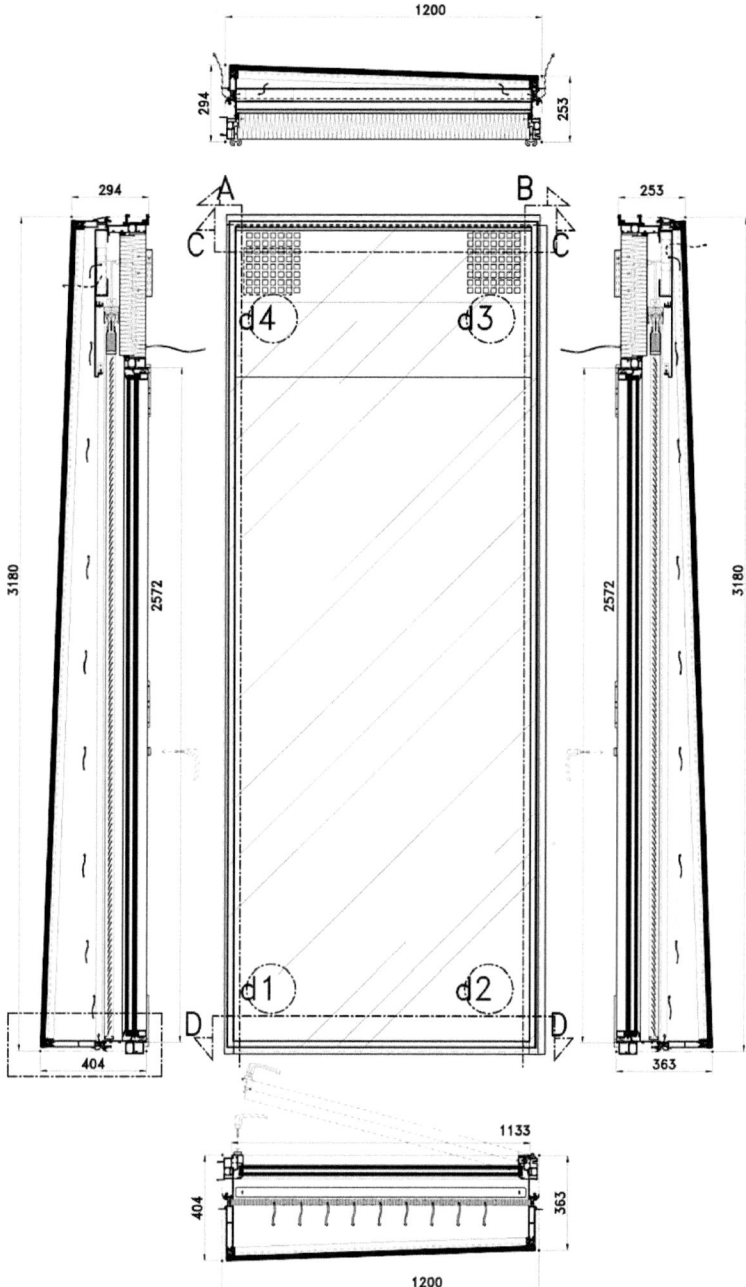

Fig. 4.29 Progetto CMR, *Garibaldi Towers*, Milan. Production drawings: modelling and functional configuration of the prefabricated unit facade component, according to the composition of the internal curtain wall, interposed cavity and external screen. © Courtesy of Progetto CMR

Fig. 4.30 Progetto CMR, *Garibaldi Towers*, Milan. Connection and interaction procedures between façade components and cladding sections, observing the wing span of the external laminated glass screens beyond the frame profiles. © Courtesy of Progetto CMR

The conception of the dynamic (and *eco-efficient*) interaction envelope system correlates with thermo-hygrometric, light and air loads, with the possibility of regulating their flows and directing them in overall functioning: in this way, the re/qualification procedures consider the fine-tuning of multi-layer unit components as mediating, diaphragmatic and osmotic devices (i.e. for reflection, capture and diffusion) between climatic conditions, typological configuration and plant systems.

The procedures for re/qualifying building envelope systems in relation to vertical types are integrated into overall functional and energy transformation strategies, through the use of certain alternative approaches to treating environmental loads (external or internal). In this regard, the strategies implemented include:

- the use of a geothermal system for heating and cooling, which uses groundwater as a thermal fluid capable of reducing carbon dioxide emissions by avoiding the use of HVAC devices;
- the adoption of photovoltaic panels, installed on the south-west façade, which contribute to energy storage, while interactive ventilation units on the façade regulate solar gain according to internal microclimatic requirements during the summer and winter seasons;
- the inclusion of interfaces for "self-regulation" in response to climatic loads, through the insertion of open perimeter spatial sections (multi-storey) which, by

limiting the impact of external horizontal loads on the main structure, accommodate the contribution of "environmental balance" systems created by bioclimatic "greenhouses" (functioning as buffer zones and sky gardens). In this case, the "greenhouses" provide increased thermal insulation by accumulating the heat required to warm the rooms in winter and cool them in summer via an appropriate ventilation system. In addition, natural cooling conditions are enhanced by the "solar chimney", which allows for energy savings of approximately 1.575 kWh/year compared to mechanical air extraction from the rooms.

Chapter 5
The Technology Transfer of the Multi-layer Façade Systems by Mass Production Components

Abstract The study examines the technical and constructive methodologies involved in the development of multiple-skin façade systems, focusing on their functional integration within architectural and environmental contexts. Through the analysis of multistorey and corridor type envelopes, the research addresses the implementation of suspended external screens (either continuous or modular) capable of generating buffer zones that mediate the relationship between internal and external climates. These configurations allow for the regulation of thermal radiation, acoustic insulation, and air circulation, supporting the maintenance of interior comfort in dynamic climatic conditions. The passive behavior of the façade system is ensured by ventilated cavities and buffer zones that respond to seasonal and daily temperature variations, facilitating vapour disposal, thermal accumulation, and internal cooling. The design integrates horizontal and vertical structural components, such as cantilevered decks, mullions, transoms, and composite frames, which are assembled with mechanical fasteners and brackets, forming a unified framework for the anchoring of curtain walls and external glass panels. The study further explores the environmental control strategies enabled by natural convection phenomena, including the solar chimney effect. This effect is achieved through the calibrated spatial relationship between the outer skin and the inner massive wall, resulting in ascending airflows driven by solar radiation and wind pressure differentials. The envelope is designed to optimize the interaction between solid and transparent elements, including large glass modules, deflector fins, and adjustable flaps, which modulate ventilation and energy exchange. The structural and morphological articulation of the technical skin contributes to the ergonomic adaptation of interior environments, enabling natural ventilation and consistent surface temperatures. Special attention is given to the integration of mass-produced and modular components tailored for flexible applications.

Keywords Suspended envelope system · Ventilated cavity design · Solar chimney effect · Thermal-acoustic regulation envelope systems · Prefabricated construction elements · Ergonomic climate mediation · Envelope mechanical fixing systems

5.1 The Technology Transfer Methods by the Transformation of Experimental and Custom Processes

The study of the multiple-skin façades systems, in the form of the multistorey or of the corridor façades, includes the application of the external screen, either continuous and homogeneous or in prefabricated modules, making it possible to determine the internal ergonomic conditions in equilibrium with respect to the climatic loads: in this case, especially to mediate the thermal and radiant flows, to reduce the heat dispersion (during the periods of reduced environmental temperature) and to reduce the acoustic loads. The application also involves the implementation of an intermediate *buffer zone* between the external and internal climate, allowing the opening of the windows referring to the main façade and maintaining the glass surfaces at a temperature close to the values of the average internal environmental temperature, so as to make the adjoining spaces more comfortable. The passive functioning of the systems provides for the ventilated cavity to perform various integrated functions (for the definition of complex mechanisms of dynamic interaction with the external environmental conditions): these are both permanent (e.g. for the increase of the thermal inertia and acoustic insulation related to the internal curtain) and temporary (e.g. for the disposal of the water vapour accumulated in the internal spaces during the periods of low environmental temperature or for the cooling of the same spaces during the periods of high environmental temperature), as exemplified by the applications on common buildings. In general, the horizontal structures perform the load-bearing surface of the horizontal framing, which provides both the contained levels between the internal curtain and the screen, and the connections for the corresponding frames: the perimeter technical skin is supported at the end of the frame by the anchoring of the mullions and transoms to the brackets.

The use of the corridor type beyond the façade sections contemplates above all the implementation of the suspended typology (structural glass façade or curtain wall) defined by fixing points with mechanical supports: the typology is aimed at constituting a functional aggregate for the environmental control at a thermos-hygrometric and acoustic level. The assembly of the envelope takes place according to the joining of the mullion profiles (integrated with each other in a specular form and provided with the hooking methods for the horizontal beams), assuming the coordination of the interface bands (and, at a performance level, of ventilation) inherent in the components realized on more levels. In general, the application is configured as follows: the screens (transparent or opaque) are framed and suspended from the substructure connected to the supporting structures or to the main perimeter sections. The dimensional calibration, opening or closing of the flaps, located at the bottom and top, can create a naturally ventilated cavity or a *buffer zone* characterized by air heating (caused by the building's thermal emission and solar radiation), leading to a reduction in heat transfer to the outside. Specifically, the air circulation is ensured by the two external flaps, one lower and one upper, which connect the external environment

Fig. 5.1 Application of a multistorey screen made of glass modules assembled to the existing curtain using aluminium frames and mechanical fixing points. © by the Authors

with the cavity; through the regulation of the opening frames and windows, it is possible to capture the loads in the cavity (Fig. 5.1).

The functioning according to the physical principles of the "solar chimney" is achieved through the space interposed between the external panel and the perimeter wall that represents the massive accumulation element. The operating principle is that the solar radiation impacts the perimeter wall and is then directed towards the external vertical enclosure: the wall increases in temperature and conveys the resulting heat to the air contained in the cavity, increasing the speed of the convective airflows. Therefore, the functioning in the "solar chimney" type is obtained as a result of the presence of the natural radiation, even with the exclusion of the shading curtain provided in an external position: the natural temperature differences between the internal spaces, the cavity and the external spaces (together with the difference in pressure caused by the wind) result in ascensional airflows. The study of the multiple-skin façades systems by the use of the external screen forming the multistorey or the corridor façades (adopting the principle of the open-joint façade that creates a more regular and continuous ventilation, by regulating the air heating-cooling cycles) concerns the processing of a series of paradigms as:

- the sustainable development to the environmental needs, the reduction of energy consumption and the application of new technologies (together with the upgrading of existing buildings). Moreover, the generation of a *heat shield*, protecting from heat through the circulation of air close to the inner temperature;

- the assembly of a diaphragm and a dynamic filter that is the expression of an architectural and technical language capable of properly satisfying the needs of well-being and comfort in sustainable manner (Fig. 5.2).

In the case study of the multiple-skin façades systems by the use of the external screen applied to *Pirelli Research and Development Centre* in Milan, designed by Gregotti Associati International, the integration is determined with respect to the composition and functionality (enveloping the 4-storey building, height = 24 m). The technical skin is incorporated into a morphological concept combining the rigourous floor plan with the tension of rooting and the spatial projection of linear structures and enclosures, as well as the perceptive and environmental interaction with the surrounding area. The architectural and structural design is defined by the unified texture of the vertical enclosures, arranged beyond the main curtain wall, arranged in an alternating sequence, upwards, by the open sections inherent in the large slots and by the thin frames that support the glass modules. The combination of the two enclosure sections, which incorporate the elevated walkways around the perimeter, creates a double-skin system aimed at establishing ergonomic conditions inside in balance with climatic load: especially to mediate thermal and radiant flows, to reduce heat loss (during periods of low temperature) and to attenuate acoustic stress. The system includes a *buffer zone* between the external and internal climate, allowing the windows and doors of the main curtain wall to be opened and the glass surfaces to be maintained at a temperature close to the average internal temperature, thus making the adjacent spaces more comfortable (Figs. 5.3 and 5.4).

The construction procedures for the technical skin are based on the cantilevered deck structure beyond the perimeter curtain wall. This is achieved through the connection of hot-dip galvanized steel brackets (with bolts connected to the halfen profile): these support the steel brackets (with through bolts) of the square section steel profiles, with the possibility of adjustment provided by horizontal slits. The upper surface of the deck structure forms the support plane for the vertical aluminium frame (in the mullions and transoms or *stick system* type), which is connected to the internal curtain wall. The section determines the primary connections of both the façade system (in the connection of the internal vertical frame and through the insertion of the lower aluminium transoms into the flap related to the enclosure of the sheet metal) and the external screen: this is supported by the vertical frame made up of composite mullions (consisting of a galvanized steel core and anodized aluminium cladding) and with a symmetrical internal structure defined by the two side sections. They allow connection to the transoms, which reproduce, in the vertical section, the box-shaped morphology in anodized aluminium, with an internal mechanism for fixing to the mullions. At the same time, the internal structure of the mullions creates the assembly field for the steel pins (in the form of glass bolt fixings), intended to suspend the monolithic glass panels (Fig. 5.5).

Fig. 5.2 Steidle + Partner, *Wacker Haus*, Munich. External screening involves installing a frame made of vertical tubular profiles supported by brackets projecting from the perimeter structures: this frame holds the opening windows by rotating the vertical pins. © Courtesy of Steidle + Partner

Fig. 5.3 Gregotti Associati International, *Pirelli Research and Development Centre*, Milan. Projection of the external curtains, marked by large horizontal slots and the frames of the glass modules: these form the perimeter sections between the two transparent curtains, for perceptive interaction and functional mediation with the external environment. © by the Authors

The perimeter surface of the deck structure on the upper levels forms the connection plane of the IPE 140 steel beams supporting the slabs, which realize the horizontal plane of the sections included in the double-skin system. This horizontal frame is integrated, crosswise and at the extremities, by two steel "U"-shaped profiles: the first has the recess facing inwards and the second has the recess facing outwards, creating a support surface for the external screen, which is bolted to the steel sheet welded to the beams. The framework (with a layer of rockwool insulation in between) supports both the lower galvanized steel sheet (bolted to the lower wings of the steel "U"-shaped profiles) and the upper aluminium sheet: at the junction with the deck structures, both sheets are fitted with wide flanges that converge with the curtain wall transoms (these are galvanized steel sheet enclosures towards both the deck and the floor and the ceiling). The external screen is assembled at the edge of the frame by galvanized steel "T"-shaped brackets (connected in continuation of the pass-through bolting already used to attach the IPE beams and the perimeter "U"-shaped steel profiles): the transverse blade of the brace supports the couple of anodized aluminium brackets (by means of bolting), whose outer wings allow the connection (by screwing) of both the mullions and the transoms (Fig. 5.6).

5.1 The Technology Transfer Methods by the Transformation ...

Fig. 5.4 Gregotti Associati International, *Pirelli Research and Development Centre*, Milan. Constitution of the façade texture consisting of cladding modules, technical glass skin (supported by fixing points to the external frame) and ventilation slots. © by the Authors

The productive and constructive constitution of the support frame for the external screen involves the typological, dimensional and connective characteristics with respect to the steel frame and the upper enclosure of the mullions (with "Ω"-shaped sheet metal in anodized aluminium): in the case of corner interfaces, assembly provides for structural connection to the metal framework using steel corner brackets; this extends into a linear profile that joins the right-angled bracket in anodized aluminium, supporting the rear structure of the mullions. In the corner interface, the mullions are coupled, with a perpendicular arrangement to support the transoms and parapet elements, or steel pins for the fixing points of monolithic glass panels.

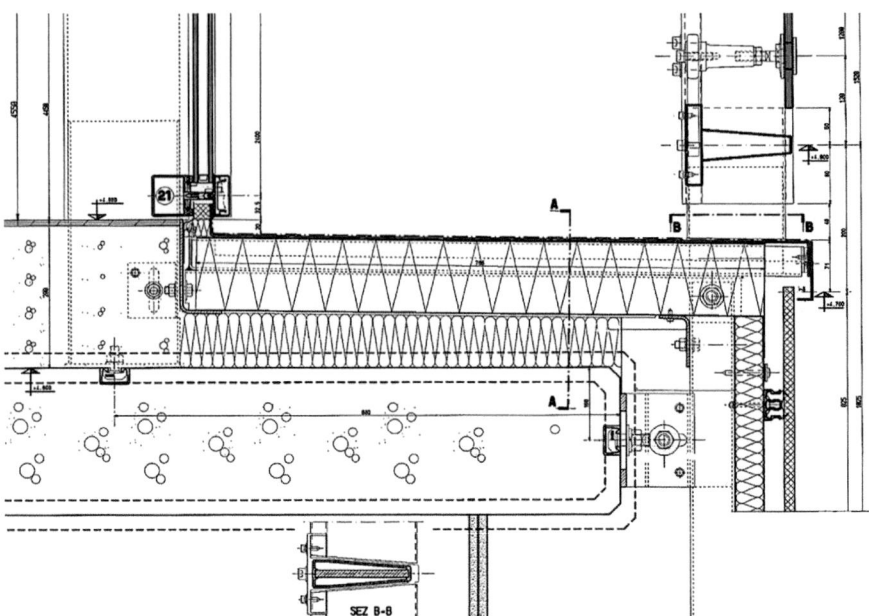

Fig. 5.5 Gregotti Associati International, *Pirelli Research and Development Centre*, Milan. Construction and technical interfaces of the double-skin system according to the mechanical connection devices (adjusted by means of brackets), layers and metal sheet fittings. © Courtesy of Focchi

Specifically, the multi-layer façade composition involves the internal curtain with aluminium profiles supporting fixed double-glazed enclosures. The combined configuration of the mullions supports the skin in glass modules, creating openings for ventilation of the space between the double envelope) (Figs. 5.7, 5.8 and 5.9).

The vertical enclosures culminate at the top with the projection of the roofing system, which is flat and extends beyond the outer curtain wall. The roof is supported by steel beams HEA 200 type, whose lower flanges create the interface conditions with respect to the mullions related to the façade system, by means of connection (by bolting) with the galvanized steel "U"-shaped bracket; the mullions for the external screen are connected to the perimeter truss by means of the connection (with upper plate) of the telescopic coupling profile. The frame is connected to the façade system by vertically inserting aluminium sheet sandwich panels (Fig. 5.10).

The application of the multiple-skin façades systems by the use of the external screen regards the upgrading and re/qualification of buildings as exemplified by the case study of the external screen assembled to *The Samaritaine* in Paris, designed by SANAA: the curved outer skin comprises cylindrical glass panels, which balance exactly on adjustable stainless steel brackets. The wide range of geometric shapes in the outer skin requires an adjustable support system that can be attached to any of the brackets connected to the building: this is possible by using a pin placed on the support that can be rotated on an axis parallel to the glass.

5.2 The Composition of Functional and Executive Principles by the Use …

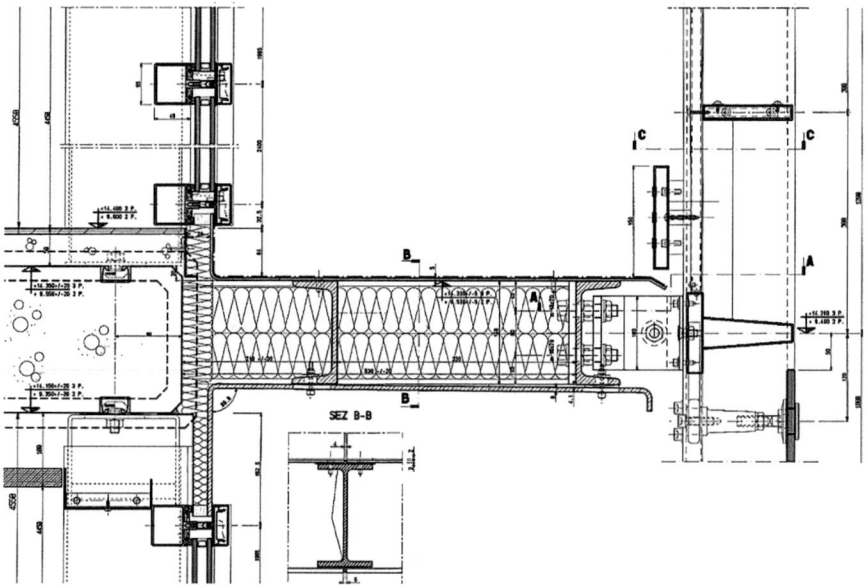

Fig. 5.6 Gregotti Associati International, *Pirelli Research and Development Centre*, Milan. Application of the horizontal load-bearing structure (as steel beams with transverse "U"-shaped profiles), which provides both the planes contained between the curtain wall and the screen, and the load-bearing joints for the corresponding frames: the perimeter skin is supported at the end of the frame by fixing the mullions and transoms to steel "T"-shaped brackets. © Courtesy of Focchi

The combination of the three glass layers means that the intermediate technical skin is equipped with a central pivot system that rotates the glass 15° on its axis, leaving space for external cleaning of the thermal skin. The wave-shaped glass façade has a running mirror print created using the magnetron process; stainless steel brackets hold each corrugated panels (with general dimensions of 2.700 × 3.500 mm) in place at four different points (Figs. 5.11 and 5.12).

5.2 The Composition of Functional and Executive Principles by the Use of Technical and Mass Production Elements

The study of the multiple-skin façades systems, with integrated and passive functioning, approaches the experimental analysis of the dynamic interaction procedures with respect to the environmental loads, external and internal to the built spaces. In this regard, the study aimed at the activation of the natural ventilation, thermo-insulating barrier effects and acoustic protection modalities is directed to the simplified, mass production, adaptable to multiple situations of use starting from tertiary,

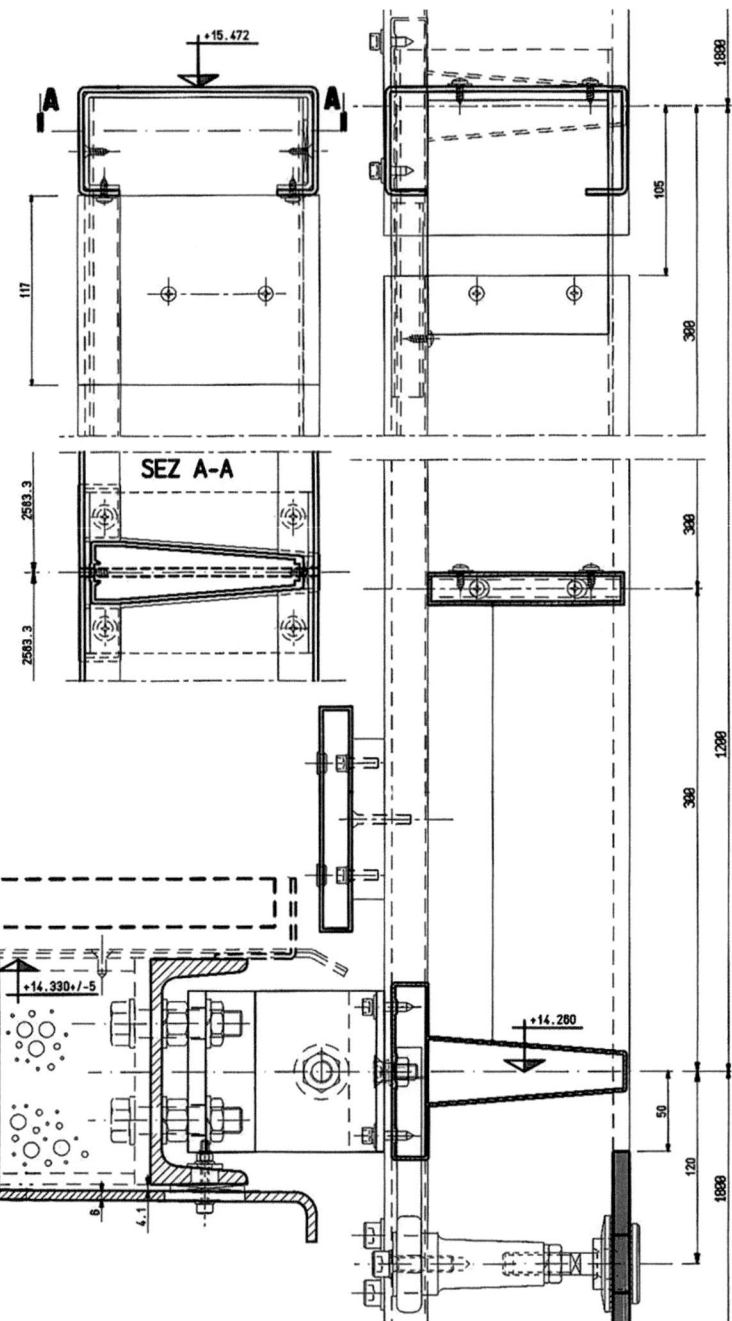

Fig. 5.7 Gregotti Associati International, *Pirelli Research and Development Centre*, Milan. Typological and structural characters of the frame supporting the external skin (in monolithic glass modules) which determines the uniformity of the perimeter surfaces, with respect to the steel frame and the upper enclosure of the mullions by the "Ω"-shaped aluminium sheet. © Courtesy of Focchi

5.2 The Composition of Functional and Executive Principles by the Use … 123

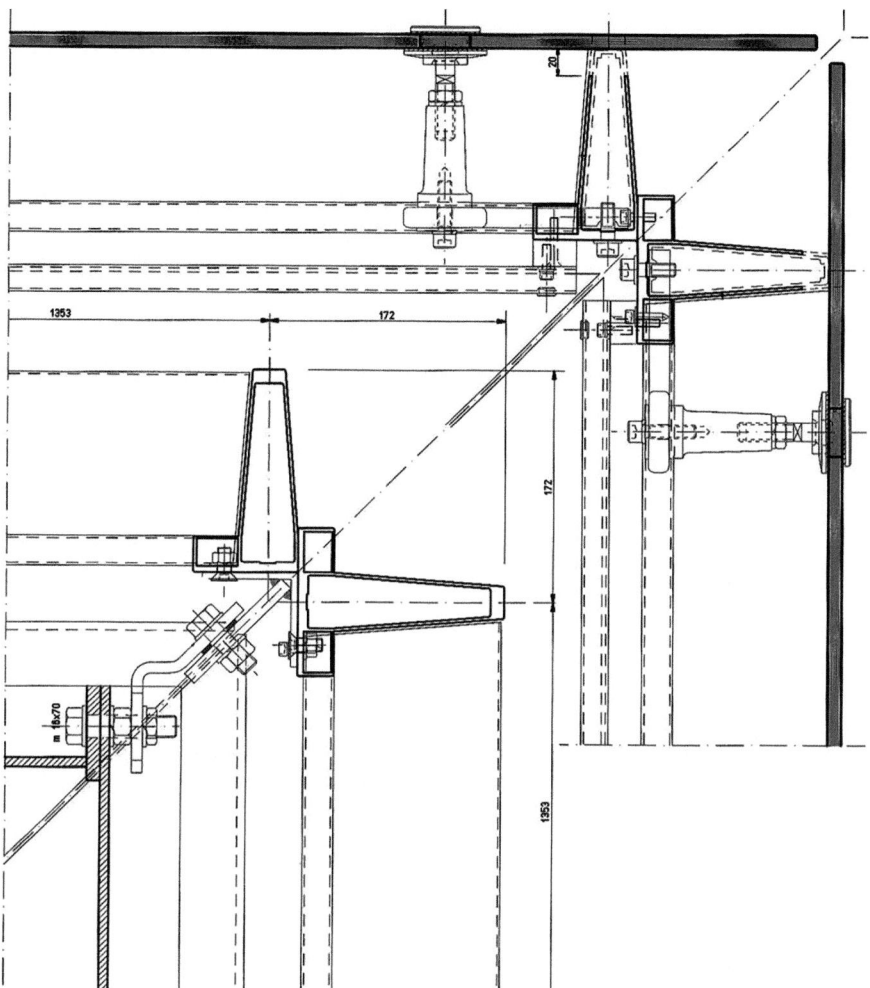

Fig. 5.8 Gregotti Associati International, *Pirelli Research and Development Centre*, Milan. Procedure for connecting linear curtains at right angles, involving the connection of the frame to the metal structure: the mullions, of the perpendicular double-profile type, support both the transoms and the steel pins for the fixing points of the monolithic glass modules. © Courtesy of Focchi

commercial and healthcare types, calibrated according to particular needs. In the case study of the multiple-skin façades systems by the use of the external screen applied to *Federal Almazov Heart, Blood and Endocrinology Centre* (within the architectural complex of the Almazova Medical Centre) in St. Petersburg, designed by Studio 44, the integration is determined according to the use of the perimeter screen (which covers the entire external surface area, equivalent to 6.000 sqm) beyond the curtain wall sections. This is achieved through the construction of a suspended façade (or *point fixed curtain wall*) defined by specific mechanical fixing points constraints:

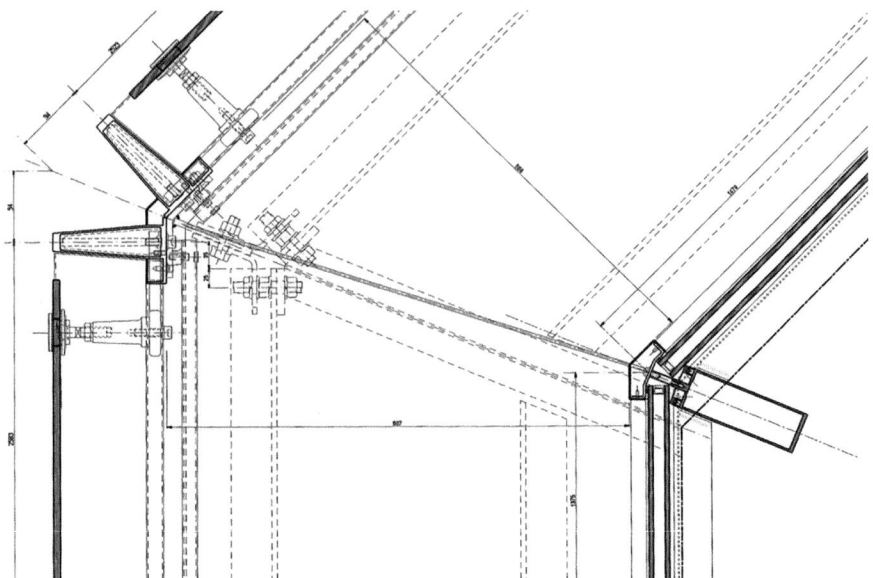

Fig. 5.9 Gregotti Associati International, *Pirelli Research and Development Centre*, Milan. Assembly procedures of the curtain wall and the modules of the external screen. © Courtesy of Focchi

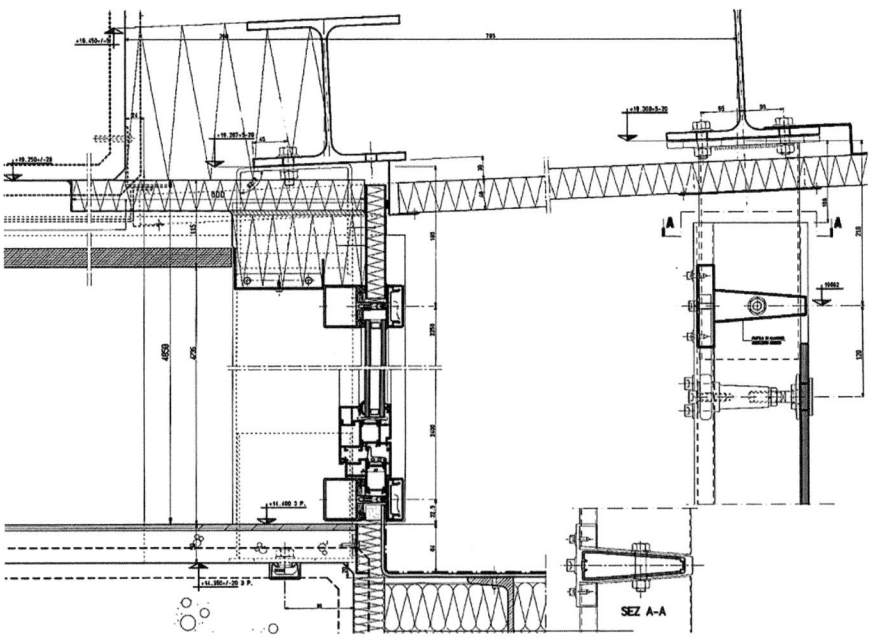

Fig. 5.10 Gregotti Associati International, *Pirelli Research and Development Centre*, Milan. Assembly procedures for the enclosure to the planar system. © Courtesy of Focchi

5.2 The Composition of Functional and Executive Principles by the Use ...

Fig. 5.11 SANAA, *Le Samaritaine*, Paris. Requalification of the building envelope by assembling curved glass curtain walls with a mechanical fastening system on adjustable stainless steel brackets (connected to the existent perimeter structure). © Courtesy of Giulia Tosarello

the type is designed to constitute a functional aggregate for environmental control management (in terms of temperature, humidity and noise), in accordance with its intended use in healthcare facilities. The calibration of the geometric and structural framework is articulated in relation to both the point-supported reinforced concrete structures and the vertical enclosures: these are arranged by inserting openings between the curtain wall, providing the external composition of the screens. The coordination procedures between the planimetric and distributive solution and both spatial and realization planning govern the calibration of the envelope texture, offering the external composition of the screen in accordance with the trace of the modules inherent to the supporting frames (able to adapt, in a flexible form, to the elliptical geometry and the objective of reducing production costs).

Fig. 5.12 SANAA, *Le Samaritaine*, Paris. Requalification of the building envelope by the external use of glass diaphragm, osmotic and a dynamic filter. © Courtesy of Giulia Tosarello

The typological processing of the envelope is combined with the requirements related to the concentric inclusion of the rooms, the distribution of vertical loads and the central location of the service systems: this leading to the structural balancing of both the planimetric and enclosure sections (through unification supported by the radial, central and peripheral concept) in accordance with the distinction of intended uses accommodated by the perspective meanings (Figs. 5.13, 5.14 and 5.15).

The pronounced execution of the horizontal structures, tapered at the extremities, provides the interface (through the inclusion of a reinforced beam) for the application of the brackets to support the screen. In addition, the processing determines the assembly works of the screen according to the joining of the mullion profiles, which are integrated into each other in a specular form, and provided with the attachment methods for the horizontal beams; the processing coordinates the interface (and, on a performance level, ventilation) slots between the components built on two levels. The technical interfaces related to the horizontal elevation structure are marked with regard to the tapered section towards the outer end, in accordance with the different dimensional configuration. The pronounced part provides for the support to the joining elements for the external screen (Figs. 5.16, 5.17 and 5.18).

5.2 The Composition of Functional and Executive Principles by the Use …

Fig. 5.13 Studio 44, *Federal Almazov Heart, Blood and Endocrinology Centre*, St. Petersburg. Building of the envelope system according to the suspended façade typology (with mechanical fixing points) for thermo-hygrometric and acoustic control. © Courtesy of Lilli Systems

The execution, beyond the structural extrados, involves the fixing of the steel plate (thickness = 3 mm, by dowelling) with the folding towards the outer perimeter (the extent of which depends on the different levels), aimed at containing the concrete screed. The connection and support structure is defined on the basis of galvanized steel plates (brackets) made over the sheet metal (directly by dowelling): these are associated by means of riveting across the structural perimeter (Fig. 5.19).

The vertical frame, consisting of the mullions made of tubular steel profiles and realized by means of a progressive vertical telescopic joint, is equipped with both the transoms and the inserts for attaching the beams made of tubular steel profiles; the frame is hooked to the brackets by means of the mechanical connection according to the shelves attached to the adjustment pin (Figs. 5.20 and 5.21).

The technical interfaces related to the roof section, proceeding according to the fine-tuning of the external projection of the floor structure (in accordance with the calibration of the thickness dimensions), consider the priority extrados application of the steel sheet. These sheet, executed in a linear form on the upper surface of the floor, bends towards the external projection in order to create support for both the flaps that cover the framing joint and the roof cladding. The processing includes both the assembly of the galvanized steel plates (brackets) on the sheet metal (by

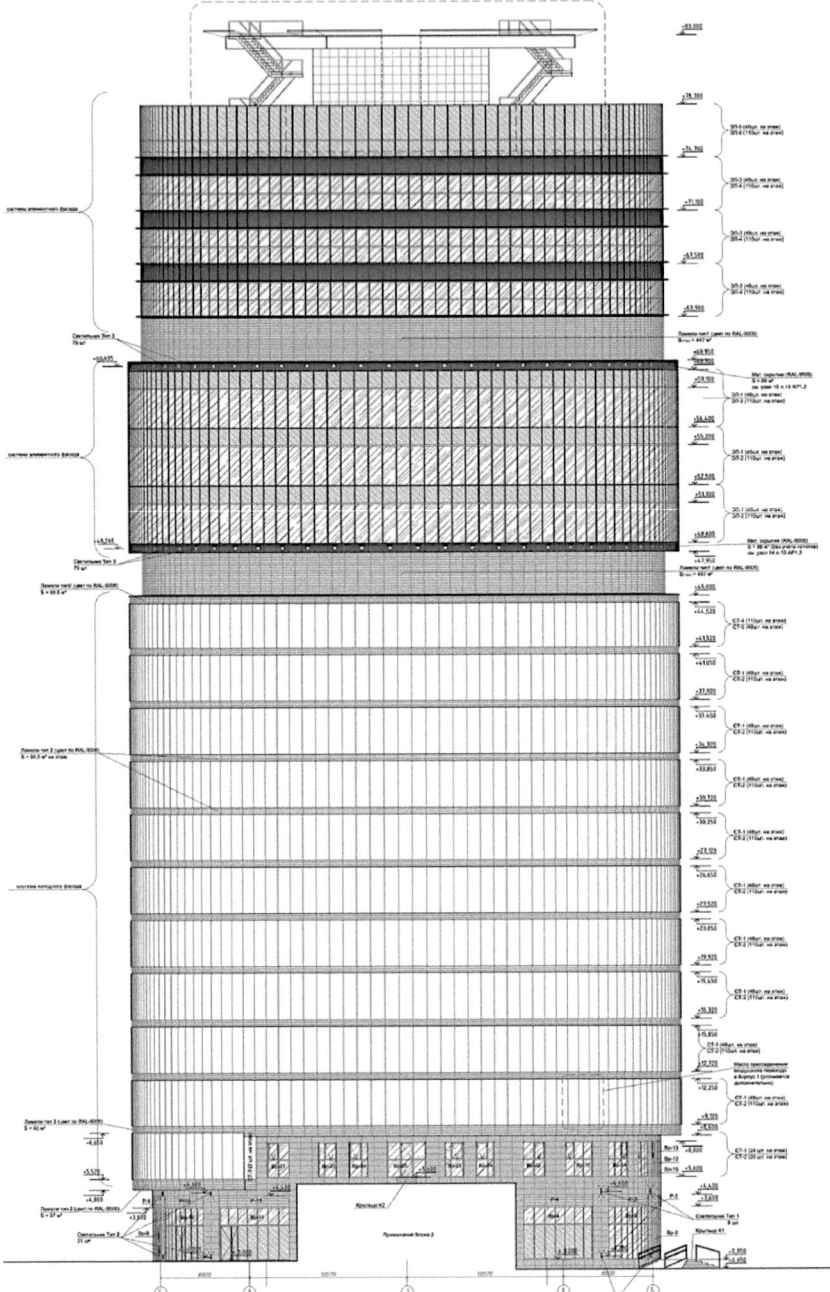

Fig. 5.14 Studio 44, *Federal Almazov Heart, Blood and Endocrinology Centre*, St. Petersburg. Executive design of both structural and enclosure modulation with respect to the sections of different use, according to the types of façade components. © Courtesy of Lilli Systems

5.2 The Composition of Functional and Executive Principles by the Use … 129

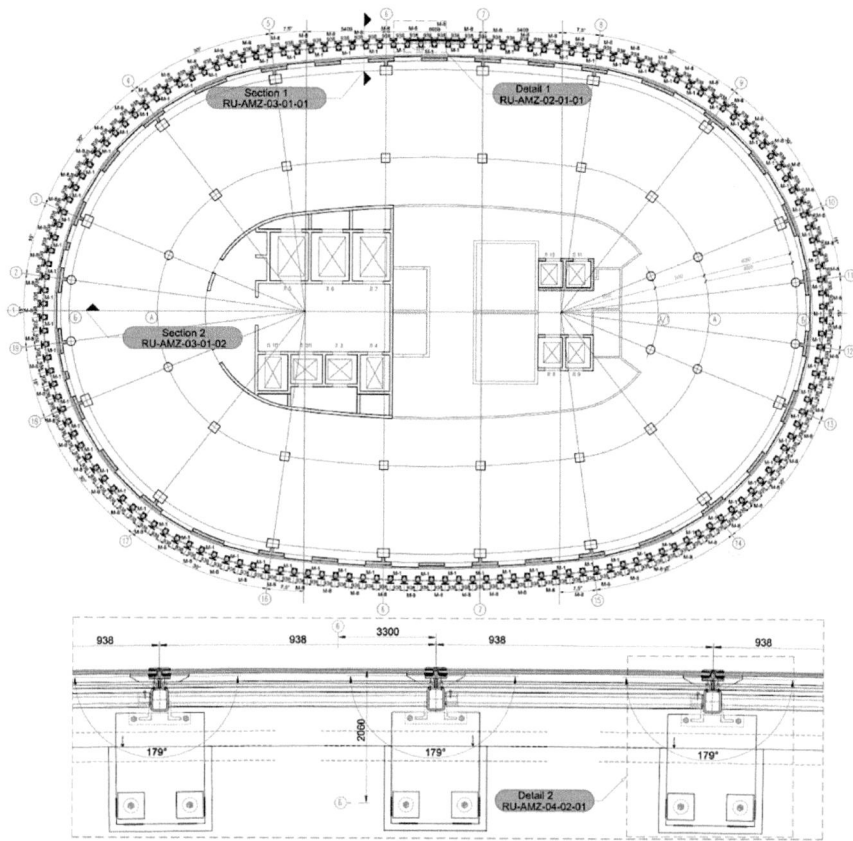

Fig. 5.15 Studio 44, *Federal Almazov Heart, Blood and Endocrinology Centre*, St. Petersburg. Executive design of the geometric, structural and operative framework concerning the connection between the typological setting and the tracing of the main and secondary load-bearing structures for the internal and external façade system. © Courtesy of Lilli Systems

dowelling) and the attachment of the cladding plates around the structural section, connected to the extrados steel plate and the inclined intrados surface. Moreover, the same steel plate forms the support plane for the "Ω"-shaped profile intended to regulate, by means of the overlapping of a "C"-shaped profile, the construction level of the roofing layers. These, starting from the heat-insulating panels (executed on the "C"-shaped profile), are arranged according to the laying of the double waterproofing sheath stretched to the steel sheet. The mullions support the external screen in glass panels (thickness = 55.2 mm) by means of extending the flanges (welded) for the assembly of the direct studs without drilling (Figs. 5.22, 5.23 and 5.24).

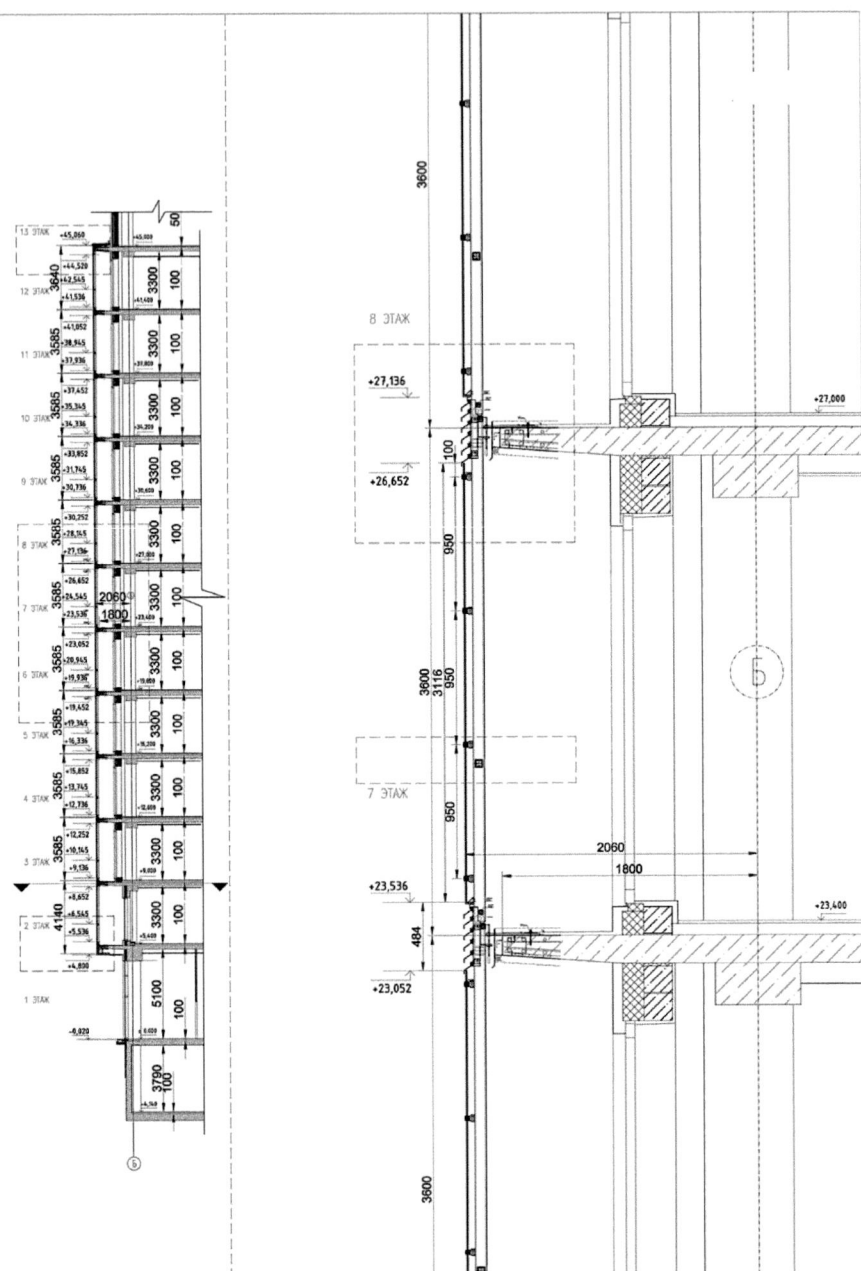

Fig. 5.16 Studio 44, *Federal Almazov Heart, Blood and Endocrinology Centre*, St. Petersburg. Executive design of the vertical and horizontal elevation structures, according to the geometric and operative coordination for the vertical enclosures and external screen. © Courtesy of Lilli Systems

5.2 The Composition of Functional and Executive Principles by the Use … 131

Fig. 5.17 Studio 44, *Federal Almazov Heart, Blood and Endocrinology Centre*, St. Petersburg. Operative coordination between the structural frame, the vertical enclosures and the secondary load-bearing devices of the external screen. © Courtesy of Lilli Systems

Fig. 5.18 Studio 44, *Federal Almazov Heart, Blood and Endocrinology Centre*, St. Petersburg. Building procedures for the frame and screen to the floor sections. © Courtesy of Lilli Systems

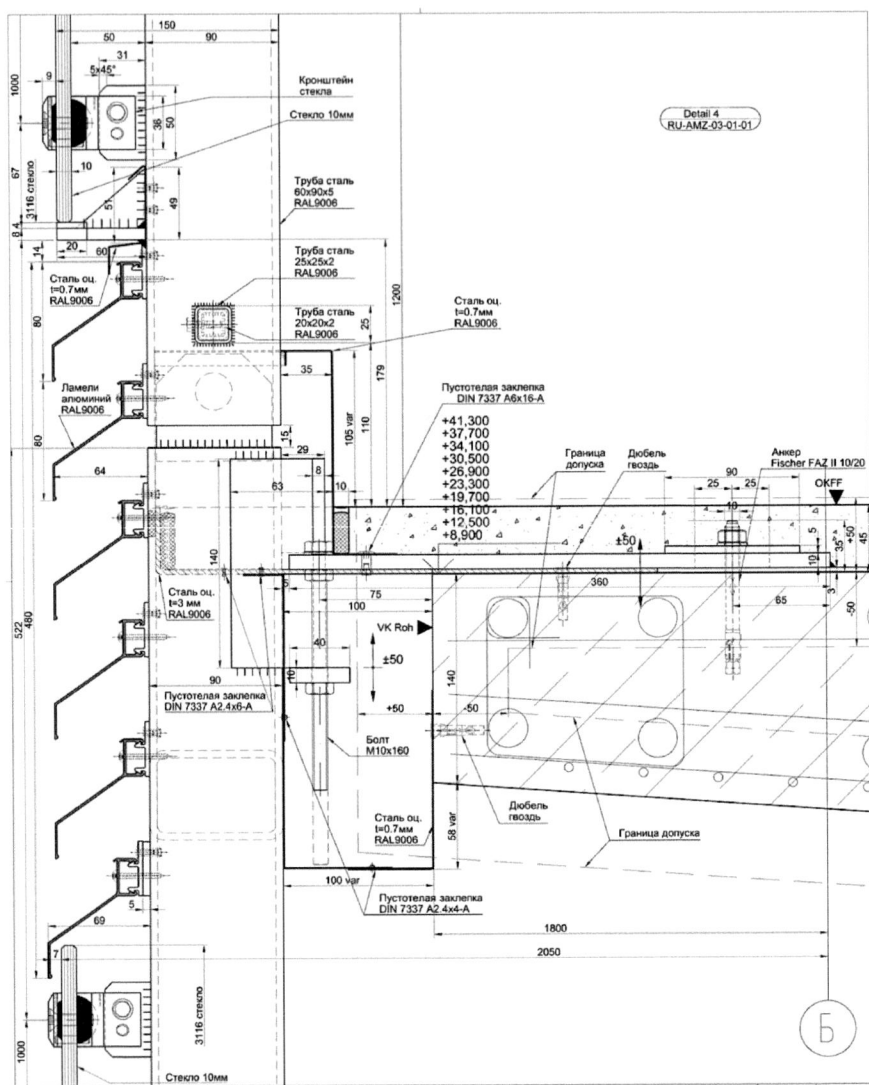

Fig. 5.19 Studio 44, *Federal Almazov Heart, Blood and Endocrinology Centre*, St. Petersburg. Processing of the technical interfaces between the horizontal elevation structure and the external screen system, according to the assembly and adjustment procedures. © Courtesy of Lilli Systems

5.2 The Composition of Functional and Executive Principles by the Use … 133

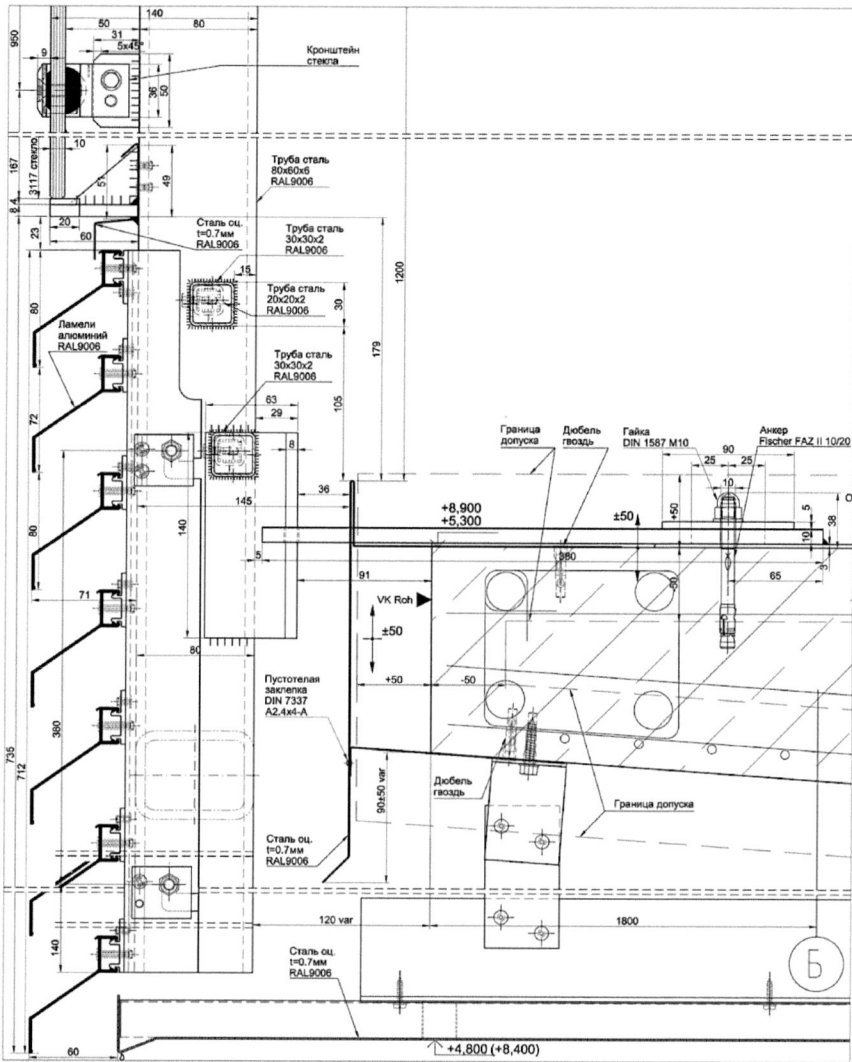

Fig. 5.20 Studio 44, *Federal Almazov Heart, Blood and Endocrinology Centre*, St. Petersburg. Processing of the technical interfaces of the external screen system according to the adjustment and connection procedures by the projections and steel sheet flaps. © Courtesy of Lilli Systems

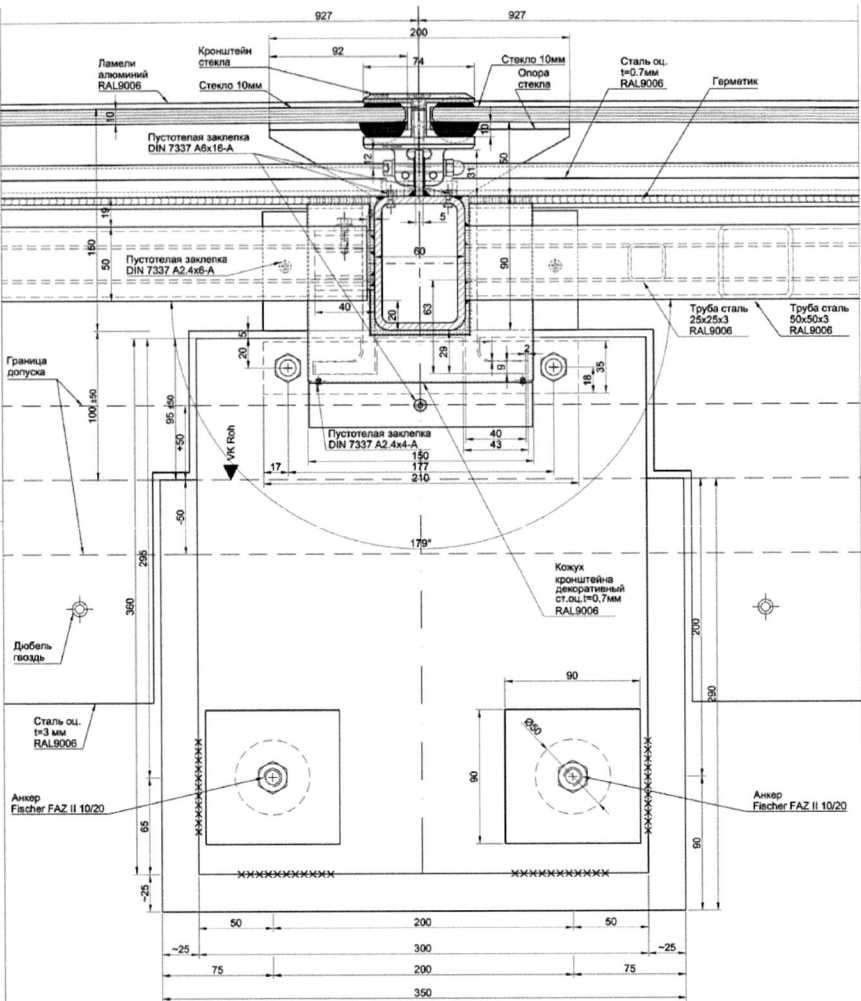

Fig. 5.21 Studio 44, *Federal Almazov Heart, Blood and Endocrinology Centre*, St. Petersburg. Processing of the technical interfaces between the horizontal elevation structure and the external screen, according to the combination of the criteria for fixing the load-bearing framework and the calibration towards the enclosure and cladding elements. © Courtesy of Lilli Systems

5.2 The Composition of Functional and Executive Principles by the Use … 135

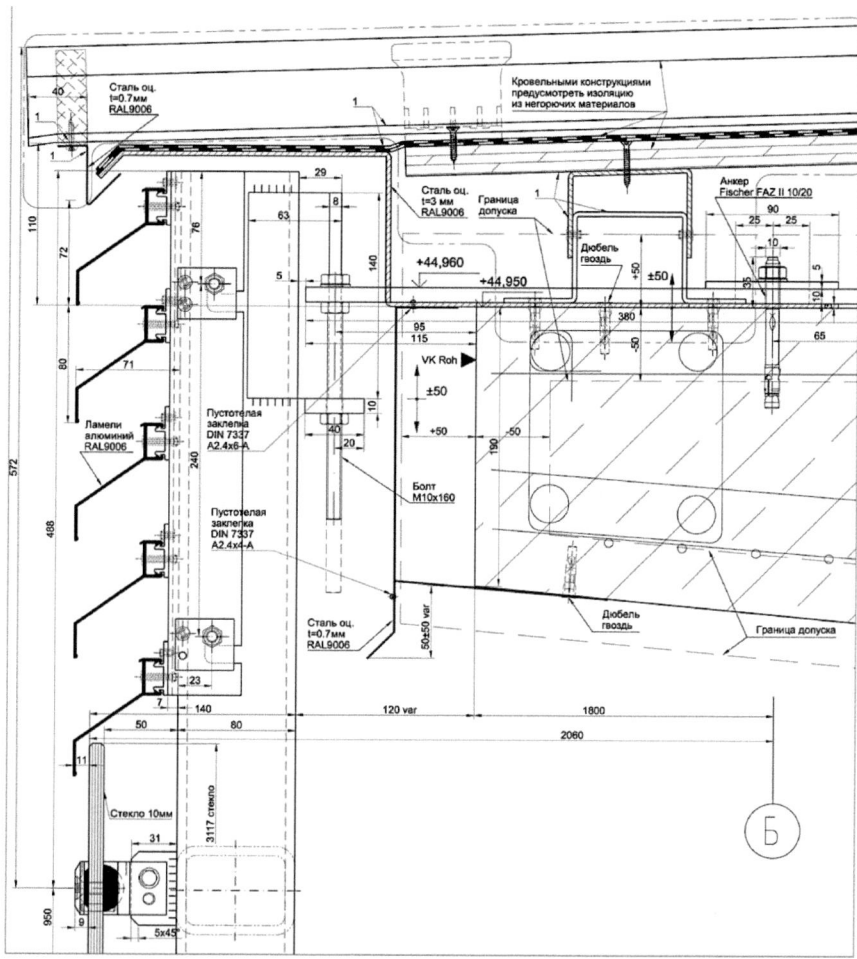

Fig. 5.22 Studio 44, *Federal Almazov Heart, Blood and Endocrinology Centre*, St. Petersburg. Processing of the technical interfaces related to the roofing section, with respect to the connection between the thermal layer, the waterproofing and the metal wrapping. © Courtesy of Lilli Systems

The corner solution of the screen system observes the adoption of the bracket with the double specular "L"-shaped reinforcements in steel to support the frame composed of the couple of tubular steel profiles: this, wrapped in the sheet metal cladding, supporting both the transoms and the tubular steel profiles, creates the bearing for the steel blade directed to the junction of the glass fasteners (Fig. 5.25). The ventilation of the cavity between the vertical enclosures and the screen is then achieved by means of the louvers established by the series of deflector fins made at the inter-floor gap level (assembled with the double septum pins) (Fig. 5.26).

Fig. 5.23 Studio 44, *Federal Almazov Heart, Blood and Endocrinology Centre*, St. Petersburg. Executive modulation between the vertical enclosures, the open sections and the external screen frame (implemented on the horizontal elevation structure). © Courtesy of Lilli Systems

Fig. 5.24 Studio 44, *Federal Almazov Heart, Blood and Endocrinology Centre*, St. Petersburg. Executive interfaces between mullions framing and external screen. © Courtesy of Lilli Systems

5.2 The Composition of Functional and Executive Principles by the Use … 137

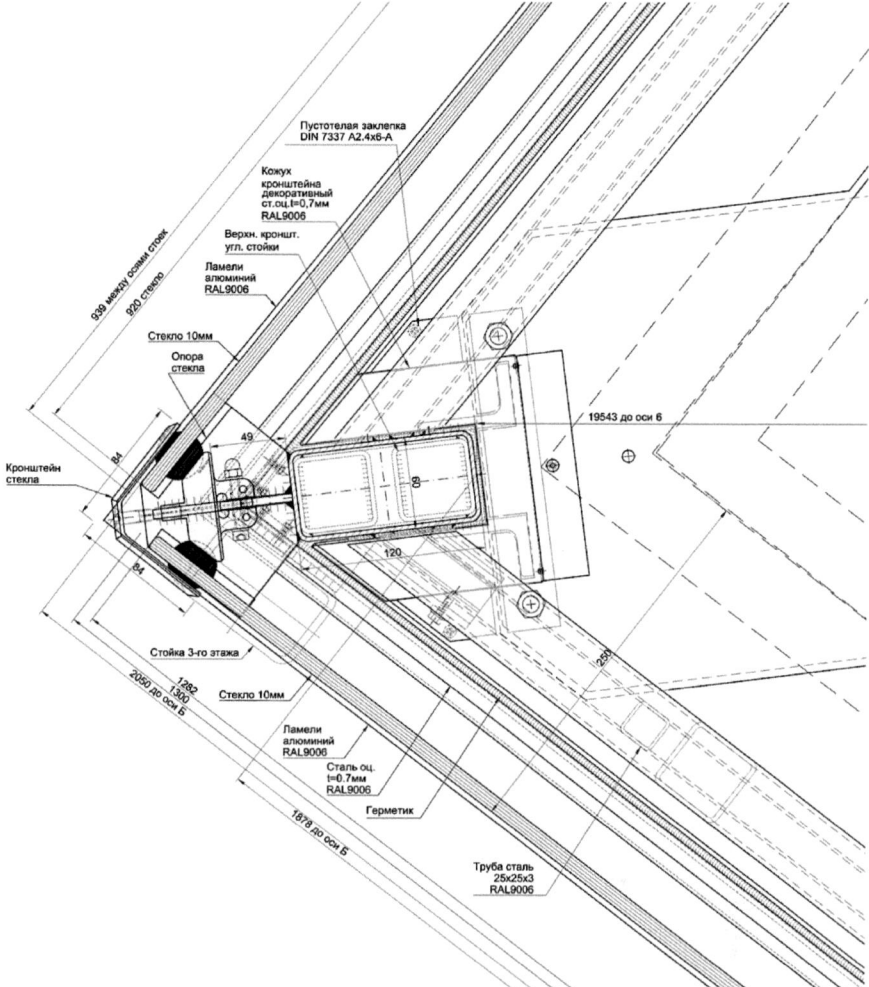

Fig. 5.25 Studio 44, *Federal Almazov Heart, Blood and Endocrinology Centre*, St. Petersburg. Executive interfaces on the angular solution of the external screen, according to the way it is joined to the double tubular profile frame and the perimeter extension. © Courtesy of Lilli Systems

The typological solution is established on the basis of the mass customized *SJS Evolution* for double-skin façades, which, by means of fixing point with a spherical joint without drilling holes in the glazing, provides adaptation to the curve of the enclosures by means of the rotation of the spherical caps (forming the joint): this eliminates the need to produce special interfaces or connection devices and ensures extremely fast installation on-site. The *SJS Evolution* type, with a load-bearing structure in aluminium profiles, with shaped brackets placed in the face of the frame and directed to the connection of the enclosures (point-supported with spherical node),

Fig. 5.26 Studio 44, *Federal Almazov Heart, Blood and Endocrinology Centre*, St. Petersburg. Implementation of ventilation slots arranged between the mullions to natural ventilate the cavity between the perimeter enclosures and the external screen. © Courtesy of Lilli Systems

offers considerable economies in the production of glass, which is not drilled, and therefore not tempered, and consequently less expensive than what is required for a *rotulle* façade. The *SJS Evolution* system is conceived according to the execution of a support and an external pressure device that, joined by means of a through-screw, provide for the addition of a spherical joint aimed at absorbing the loads induced both by external actions and by the deformations of the horizontal structures, by means of the blocking determined by two spherical caps made of plastic material with a low friction coefficient: these are fitted in two spherical insertions on both the support and the external pressure device (with the same diameter as the sphere obtained from the joining). The absence of holes reduces fabrication costs and assembly time, allowing the use of laminated glass panels (Figs. 5.27 and 5.28).

5.2 The Composition of Functional and Executive Principles by the Use …

Fig. 5.27 Studio 44, *Federal Almazov Heart, Blood and Endocrinology Centre*, St. Petersburg. Application of mass customized fixing point devices and retaining bracket to support the glass screen (without drilling) in respect of the mullion frame. © Courtesy of Lilli Systems

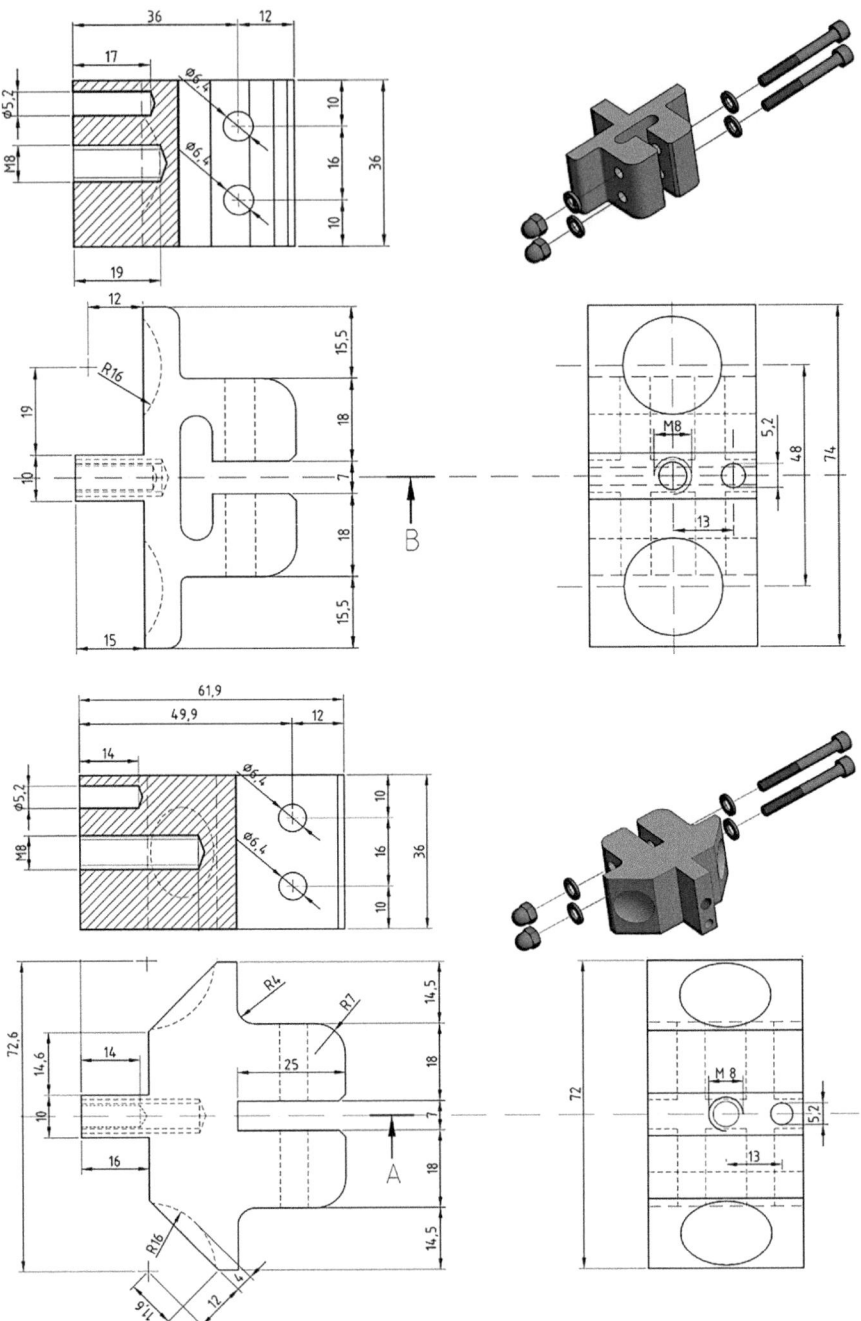

Fig. 5.28 Studio 44, *Federal Almazov Heart, Blood and Endocrinology Centre*, St. Petersburg. Production processing of the fixing points of glass screens: coupling devices to the block